Revise
GCSE

Physics

Contents

This book and

	AQA	**Edexcel**
Web Address	www.aqa.org.uk	www.edexcel.com
Specification Number	4403	2PH01
Exam Assessed Units and Modules At least 40% of assessment must be carried out at the end of the course. For students starting the GCSE course from September 2012 onwards, all assessment (100%) must take place at the end of the course.	**Unit 1: Physics 1** 1 hr 60 marks 25% of GCSE **Unit 2: Physics 2** 1 hr 60 marks 25% of GCSE **Unit 3: Physics 3** 1 hr 60 marks 25% of GCSE All papers feature structured and closed questions.	**Unit P1** 1 hr 60 marks 25% of GCSE **Unit P2** 1 hr 60 marks 25% of GCSE **Unit P3** 1 hr 60 marks 25% of GCSE All papers feature objective, short answer and extended writing questions.
Controlled Assessment Covering: · Research, planning and risk assessment · Data collection · Processing, analysis and evaluation	**Unit 4: Controlled Assessment** 1hr 35 min, plus time for research/data collection 50 marks 25% of GCSE	**Unit PCA** Approx. 3 hrs 50 marks 25% of GCSE
Chapter Map*		
1 Energy	P1.1, P1.2, P1.3, P1.4, P3.3	P1, P2
2 Waves	P1.5	P1, P3
3 Electromagnetic waves	P1.1, P1.5, P3.1	P1, P3
4 Beyond the Earth	P1.5, P2.6	P1
5 Electricity	P2.3, P2.4	P1, P2
6 Forces and motion	P1.2, P2.1, P2.2	P1, P2, P3
7 Radioactivity	P2.5, P2.6, P3.1	P2, P3
8 Light	P3.1	P1, P3
9 Further physics	P1.1, P3.1, P3.2, P3.3	P1, P3

* There are tick charts throughout the book to show which particular sub-topics in each chapter are relevant to your course.

your GCSE course

OCR A	OCR B	WJEC	CCEA
www.ocr.org.uk	www.ocr.org.uk	www.wjec.co.uk	www.ccea.org.uk
J245	J265	600/1032/0	G76
Modules P1, P2 and P3 1 hr 60 marks 25% of GCSE **Modules P4, P5 and P6** 1 hr 60 marks 25% of GCSE **Module P7** 1 hr 60 marks 25% of GCSE All papers feature objective style and free response questions.	**Modules P1, P2 and P3** 1hr 15 min 75 marks 35% of GCSE **Modules P4, P5 and P6** 1hr 30 min 85 marks 40% of GCSE Includes a data response section worth 10 marks, which assesses AO3 All papers feature structured questions.	**Physics 1** 1 hr 60 marks 25% of GCSE **Physics 2** 1 hr 60 marks 25% of GCSE **Physics 3** 1 hr 60 marks 25% of GCSE All papers feature structured questions involving some extended writing.	**Unit P1** Higher: 1 hr 30 min 100 marks Foundation: 1 hr 15 min 80 marks 35% of GCSE **Unit P2** Higher: 1 hr 45 min 115 marks Foundation: 1 hr 30 min 90 marks 40% of GCSE All papers feature structured questions.
Unit A184 Approx. 4.5–6hrs 64 marks 25% of GCSE	**Unit B753** Approx. 7 hrs 48 marks 25% of GCSE	**Unit CA** Approx. 7.5 hrs, plus time for initial research 48 marks 25% of GCSE	**Unit 3** 1 hr, plus time for planning, risk assessment and data collection 45 marks 25% of GCSE
P3, P5	P1, P2, P4, P6	P1, P3	P1, P2
P1, P7	P1, P4, P5	P1, P3	P2
P2, P7	P1, P2, P4, P5	P1	P1, P2
P1, P7	P1, P2, P5	P1, P3	P2
P5	P4, P6	P2, P3	P2
P4	P3, P5	P2, P3	P1
P6, P7	P2, P4	P1, P2, P3	P1
P7	P1, P5	P3	P2
P5	P2, P4, P5, P6	P3	P1, P2

Preparing for the exams

What will be assessed

In your science exams and controlled assessment, you are assessed on three main criteria called assessment objectives:

- **Assessment Objective 1 (AO1)** – tests your ability to **recall**, select and communicate your knowledge and understanding of physics
- **Assessment Objective 2 (AO2)** – tests your ability to **apply** your skills, knowledge and understanding of physics in practical and other contexts
- **Assessment Objective 3 (AO3)** – tests your ability to **analyse** and **evaluate** evidence, make reasoned judgements and draw conclusions based on evidence

The exam papers have a lot of AO1 and AO2 questions and some AO3 questions. The controlled assessments focus mainly on AO2 and AO3.

To do well on the exams, it is not enough just to be able to recall facts. You must be able to apply your knowledge to different scenarios, analyse and evaluate evidence and formulate your own ideas and conclusions.

Planning your study

It is important to have an organised approach to study and revision throughout the course.

- After completing a topic in school or college, go through the topic again using this guide. Copy out the main points on a piece of paper or use a pen to highlight them.
- Much of memory is visual. Make sure your notes are laid out in a logical way using colour, charts, diagrams and symbols to present information in a visual way. If your notes are easy to read and attractive to the eye, they will be easier to remember.
- A couple of days later, try writing out the key points from memory. Check differences between what you wrote originally and what you wrote later.
- If you have written your notes on a piece of paper, make sure you keep them for revision later.
- Try some of the questions in this book and check your answers.
- Decide whether you have fully mastered the topic and write down any areas of weakness you think you have.

How this book will help you

This complete study and revision guide will help you because...

- it contains the essential content for your GCSE course without any extra material that will not be examined
- there are regular short progress checks so that you can test your understanding
- it contains sample GCSE questions with model answers and notes, so that you can see what the examiner is looking for.
- it contains exam practice questions so that you can confirm your understanding and practise answering exam-style questions
- the summary table on pages 4-5, and the exam-board tick charts throughout the book, will ensure that you only study and revise topics that are relevant to your course.

Six ways to improve your grade

1. Read the question carefully

Many students fail to answer the actual question set. Perhaps they misread the question or answer a similar question that they have seen before. Read the question once right through and then again more slowly. Underline key words in the question as you read through it. Questions at GCSE often contain a lot of information. You should be concerned if you are not using the information in your answer.

Take notice of the command words used in questions and make sure you answer appropriately:

- **State:** A concise, factual answer with no description or explanation
- **Describe:** A detailed answer that demonstrates knowledge of the facts about the topic
- **Explain:** A more detailed answer than a description; give reasons and use connectives like 'because'.
- **Calculate:** Give a numerical answer, including working and correct units
- **Suggest:** A personal response supported by facts.

2. Give enough detail

If a part of a question is worth three marks, you should make at least three separate points. Be careful that you do not make the same point three times, but worded in a slightly different way. Draw diagrams with a ruler and label with straight lines.

3. Be specific

Avoid using the word 'it' in your answers. Writing out in full what you are referring to will ensure the examiner knows what you are talking about. This is especially important in questions where you have to compare two or more things.

4. Use scientific language correctly

Try to use the correct scientific language in your answers. The way scientific language is used is often the difference between successful and unsuccessful answers. As you revise, make a list of scientific terms you come across and check that you understand what they mean. Learn all the definitions. These are easy marks and they reward effort and good preparation.

5. Show your working

All science papers include calculations. Learn a set method for solving a calculation and use that method. You should always show your working in full. That way, if you make an arithmetical mistake, you may still receive marks for applying the correct science. Check your answer is given to the correct level of accuracy (significant figures or decimal places) and give the correct units.

6. Brush up on your writing skills

Your exam papers will include specific questions for which the answers will be marked on both scientific accuracy and the quality of the written communication. These questions are worth 6 marks, but it does not matter how good the science is, your answer will not gain full marks unless:

- the text is legible and the spelling, punctuation and grammar are accurate so that your meaning is clear
- you have used a form and style of writing that is fit for purpose and appropriate to the subject matter
- you have organised information in a clear and logical way, correctly using scientific vocabulary where appropriate.

These questions will be clearly indicated on the exam papers.

Exam papers are scanned and marked on a computer screen. Do not write outside the answer spaces allowed, or your work may not be seen by the examiner. Ask for extra paper if you need it. Choose a black pen that will show up – one that photocopies well is a good choice.

How Science Works

The science GCSE courses are designed to help develop your knowledge of certain factual details, but also your understanding of 'How Science Works'.

'How Science Works' is essentially a set of key concepts that are relevant to all areas of science. It is concerned with four main areas:

Data, evidence, theories and explanations

- science as an evidence-based discipline
- the collaborative nature of science as a discipline and the way new scientific knowledge is validated
- how scientific understanding and theories develop
- the limitations of science
- how and why decisions about science and technology are made
- the use of modelling, including mathematical modelling, to explain aspects of science

Practical skills

- developing hypotheses
- planning practical ways to test hypotheses
- the importance of working accurately and safely
- identifying hazards and assessing risks
- collecting, processing, analysing and interpreting primary and secondary data
- reviewing methodology to assess fitness for purpose
- reviewing hypotheses in light of outcomes

Communication skills

- communicating scientific information using scientific, technical and mathematical language, conventions and symbols.
- use models to explain systems, processes and abstract ideas

Applications and implications of science

- the ethical implications of physics and its applications
- risk factors and risk assessment in the context of potential benefit

You will be taught about 'How Science Works' throughout the course in combination with the scientific content. Likewise, the different exam boards have included material about 'How Science Works' in different parts of their assessment.

'How Science Works' will be assessed in the controlled assessment, but you will also get questions that relate to it in the exams. If you come across questions about unfamiliar situations in the exam, do not panic and think that you have not learnt the work. Most of these questions are designed to test your skills and understanding of 'How Science Works', not your memory. The examiners want you to demonstrate what you know, understand and can do.

1 Energy

The following topics are covered in this chapter:

- **The electricity supply**
- **Generating electricity**
- **Renewable sources of energy**
- **Electrical energy and power**
- **Electricity matters**
- **Particles and heat transfer**

1.1 The electricity supply

LEARNING SUMMARY

After studying this section, you should be able to:

- Explain what is meant by sustainable energy.
- Explain the difference between a primary energy source and a secondary energy source.
- Calculate the efficiency of electrical appliances and power stations.
- Draw Sankey diagrams to show energy transfers.

Energy resources

AQA	P1	✓
OCR A	P3	✓
OCR B	P1	✓
EDEXCEL	P1	✓
WJEC	P1	✓
CCEA	P1	✓

People in modern societies use a lot of **energy resources**. Energy resources we use directly are called **primary energy resources**. Examples include diesel, petrol and solar energy. **Electricity** is a **secondary energy resource**. It has to be made from a primary resource. We use a lot of electricity because:

- It's easy to transmit long distances.
- It can be used to do lots of different jobs.
- There is no pollution at the point where it is used.

> **KEY POINT**
>
> A **sustainable energy** supply is one that meets our needs without leaving problems for future generations, for example, without using up all the resources, causing air pollution or climate change.

Efficiency

AQA	P1	✓
OCR A	P3	✓
OCR B	P1, P2	✓
EDEXCEL	P1	✓
WJEC	P1	✓
CCEA	P1	✓

Whenever energy is transferred from a primary resource to electricity, some of it is wasted as heat in the process. No power station is **100% efficient**. In a coal-fired power station, for example, for every 100 J of energy stored in the coal that is burned, only 40 J is transferred to the electricity. Energy is a conserved quantity, which means that the total amount of energy remains the same. The 60 J of energy that is not transferred to the electricity is wasted and ends up as heat in the surrounding environment. This will cause a slight temperature increase in the surroundings.

> **KEY POINT**
>
> The efficiency of an energy transfer is the fraction, or percentage, of the energy input that is transferred to useful energy output.
>
> $$\text{Efficiency} = \frac{\text{useful energy output}}{\text{total energy input}}$$
>
> Or as a percentage:
>
> $$\text{Efficiency} = \frac{\text{useful energy output}}{\text{total energy input}} \times 100\%$$

Draw two Sankey diagrams in your notes, one for a process that has more wasted energy than useful energy and one with more useful energy than wasted energy. Label the input energy, which is 100%, and the useful energy and wasted energy. When you look at them, they will remind you of all the information summarised by a Sankey diagram.

Energy transfers taking place in a process can be shown on a **Sankey energy flow diagram**.

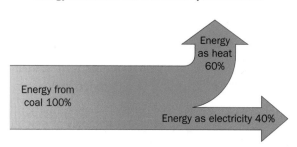

Energy transformed in a coal-fired power station.

Energy as heat 60%

Energy from coal 100%

Energy as electricity 40%

The width of the arrows is proportional to the amount of energy represented by the arrow.

Improving efficiency

AQA	P1	✓
OCR A	P3	✓
OCR B	P1, P2	✓
EDEXCEL	P1	✓
WJEC	P1	✓

The European Union (EU) has banned the sale of 100 W filament lamps because about 90% of the input energy is wasted as heat. Their efficiency is only 10% because only 10% of the output is useful light. Replacement lamps like compact fluorescent lamps (CFLs) are more efficient. For example, a CFL that uses 18 J of energy each second emits 10 J of light.

Its efficiency $= \dfrac{10}{18} \times 100 = 56\%$

A standard 100 W filament lamp

A fluorescent lightbulb

To get an A* you should be able to compare the efficiency of different processes and explain that it is important to develop more efficient ways of generating electricity to save energy resources.

Gas-fired power stations are more **efficient** than coal-fired power stations. More of the chemical energy stored in the gas is transferred to the electricity. Gas-fired power stations are about 50% efficient, whereas coal-fired power stations are about 40% efficient.

Solar energy

AQA	P1	✓
OCR A	P3	✓
OCR B	P2	✓
WJEC	P1	✓

There are two ways of using solar energy:

- **Passive solar heating** uses radiation from the Sun to heat water. Solar panels may contain water that is heated by radiation from the Sun. This water may then be used to heat buildings or provide domestic hot water.
- **Solar cells**, or **photocells**, transfer energy from sunlight into electricity.

Solar cells

| OCR B | P2 | ✓ |
| WJEC | P1 | ✓ |

Solar cells contain crystals of **silicon**. Light gives the silicon atoms energy and knocks **electrons** loose from the atoms. The electrons flow as an **electric current**. The current (which is D.C, see page 20) is increased by increasing:

- The surface area that the light falls on.
- The light intensity.

PROGRESS CHECK

1. Electricity is a secondary energy resource. Explain what this means.
2. A power station burns natural gas. Every second 755 MJ of electricity is generated from 1300 MJ of energy from the gas.
   a) Draw a Sankey diagram for the process.
   b) Calculate the efficiency of the power station.
3. An electric light uses 60 J of electrical energy every second and produces 8 J of light energy each second. Calculate its efficiency.
4. A power station burns coal and converts 3500 kJ of chemical energy to electrical energy every second. 2010 kJ of energy is wasted as heat in the power station and 300 kJ is wasted in the overhead cables every second.
   a) Draw a Sankey diagram.
   b) Calculate the efficiency of the power station in delivering useful electricity to your home.
5. Explain why it is more efficient to use a primary energy resource to heat your home than to use an electric heater.

5. When the primary energy resource is used to generate electricity, energy is wasted as heat in the power station. If the primary resource is burned in your home all the heat is useful.
4. a) 3500 kJ = 100% on left; useful = 1190 kJ = 34%; two wasted arrows; heat in power station = 2010 kJ = 57%; overhead cables = 300 kJ = 9%
   b) (1190 ÷ 3500) × 100 = 34%
3. 13%
2. a) Arrow 100% = 1300 MJ on left; useful energy = 755 MJ; wasted energy = 545 MJ
   b) (755 ÷ 1300) × 100 =58%
1. It has to be made from another, primary, energy source

1.2 Generating electricity

After studying this section, you should be able to:

- Draw a flow diagram for a power station.
- Explain the steps in generating electricity in power stations.
- Explain the process of electromagnetic induction.
- List the fuels that are used in power stations.

Power stations

AQA	P1	✓
OCR A	P3	✓
OCR B	P2	✓
EDEXCEL	P1	✓
WJEC	P1	✓

Turning a **generator** produces electricity. To turn the generators we connect them to **turbines**. We use different energy resources to turn the turbines. Wind and water flow can turn turbines directly. Steam is often used, produced by heating water. The heating is done by burning fuels or using other heat sources. The diagram shows the parts of a coal-fired power station. In a modern **gas-fired** power station the hot exhaust gases from the burners are used to turn the turbines and then to heat water to steam which turns the turbines.

A coal-fired power station.

Learn to label the different parts of a power station and practise by drawing a flow diagram.
1 Furnace (or nuclear reactor)
2 Boiler
3 Turbine
4 Generator
5 Transformer

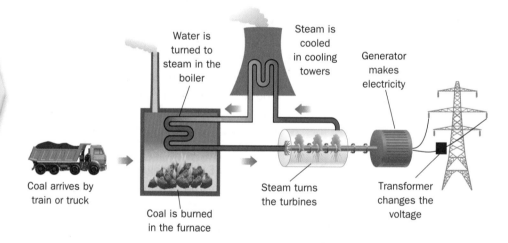

Water is turned to steam in the boiler

Steam is cooled in cooling towers

Generator makes electricity

Coal arrives by train or truck

Coal is burned in the furnace

Steam turns the turbines

Transformer changes the voltage

The main fuels

AQA	P1	✓
OCR A	P3	✓
OCR B	P2	✓
EDEXCEL	P1	✓
WJEC	P1	✓
CCEA	P1	✓

Electricity can be generated in large power stations from different fuels:

- **Fossil fuels** such as **coal**, **natural gas** and **oil**. Fossil fuels were formed over 300 million years ago. They take millions of years to form so that they will eventually run out. All the fossil fuels produce carbon dioxide when burned. Carbon dioxide absorbs infrared radiation and warms the atmosphere. The extra carbon dioxide from burning fossil fuels may be the cause of global warming and could cause climate change.

For AQA you need to know about carbon capture technology – ways of safely storing the carbon dioxide.

- In a **nuclear** reaction a large amount of energy is released from a small amount of **plutonium** or **uranium** by fission (see page 21). One advantage is that no carbon dioxide is formed.
- **Biofuels** such as **wood**, **sugar**, **straw** and **manure**. Biofuels are materials from recently living plants and animals. Carbon dioxide is removed from the

atmosphere when plants grow and then released when fuels are burned or fermented. As long as more plants are grown there is no net change. So biofuels are carbon neutral.

Lifecycle assessments

OCR A	P3	✓
WJEC	P1	✓
CCEA	P1	✓

When we decide which fuels to use in power stations we have to take lots of factors into account:

- The cost of the fuel.
- The availability of the fuel – will it run out or will it be difficult to obtain?
- The start-up time – how long it will take to commission (plan and build)?
- The cost of building the power station.
- How much it will cost to decommission (take apart at the end of its life) and to dispose of waste materials.
- The maintenance costs.
- The pollution and waste, for example whether carbon dioxide is emitted.
- The risk of accidents.

> Make sure you can compare advantages and disadvantages of the different fuels on page 12 and the renewables in the next section.

Generators

OCR A	P3, P5	✓
OCR B	P2, P6	✓
EDEXCEL	P1	✓
WJEC	P3	✓
CCEA	P2	✓

> **KEY POINT**
>
> A voltage is **induced** across a coil of wire by moving a magnet into or out of a coil. Moving the coil instead of the magnet would have the same effect.
>
> This process is called **electromagnetic induction**.

A voltage is induced.

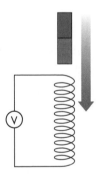

This effect is used in **dynamos** and **generators**. There are several ways to increase the voltage:

- Use stronger magnets.
- Use more turns of wire in the coil.
- Move the magnet (or the coil) faster.
- Place an iron core inside the electromagnetic coil.

A generator in a power station uses an electromagnet to produce a magnetic field. The electromagnet rotates inside coils of wire so that the electromagnet coils are in a changing magnetic field and a voltage is induced.

The bigger the voltage, the bigger the current, so to supply a large current the generator will use more primary fuel each second.

Electromagnetic induction

AQA	P3	✓
OCR A	P3, P5	✓
OCR B	P2, P6	✓
EDEXCEL	P1	✓
WJEC	P3	✓
CCEA	P2	✓

A voltage is only induced when there is movement of the magnet or coil. The direction of the voltage is reversed when the movement is reversed, or when the poles of the magnet are swapped.

PROGRESS CHECK

1. What is the difference between the way the energy is produced from the source in a coal-fired power station and a nuclear power station?
2. What is the energy source in a nuclear power station?
3. Which of these are biofuels: chicken manure, coal, natural gas, uranium or willow?
4. Give two advantages and two disadvantages of a biofuel power station.

1. Coal is burned but uranium/plutonium is split in a nuclear reaction (fission).
2. Uranium or plutonium.
3. Chicken manure and willow.
4. Examples of advantages: carbon dioxide released is matched by carbon dioxide taken up during life of fuel, so there is no net change. A way of using up waste products like manure, or methane.
Disadvantages: large amount of land required to grow some crops – maybe conflict with food crops or wildlife. Supply needs to be planned – e.g. willow planted.

1.3 Renewable sources of energy

LEARNING SUMMARY

After studying this section, you should be able to:
- Explain what a renewable source of energy is.
- List the renewable energy sources.
- Give advantages and disadvantages of renewable energy sources.
- Compare renewable and non-renewable energy sources.

Renewable resources

AQA	P1	✓
OCR A	P3	✓
OCR B	P2	✓
EDEXCEL	P1	✓
WJEC	P1	✓
CCEA	P1	✓

KEY POINT

Renewable sources of energy are those that are being made today and so will not be used up. Most renewable resources make use of the Sun's energy.

The Sun evaporates water and causes the rain that fills the rivers. The Sun also causes convection currents that produce winds, which produce waves. **Geothermal** heat and tidal energy do not originate from the Sun's energy. Geothermal heat originates from inside the Earth and the tides are caused by the Moon.

These are the renewable resources:

To remember all the renewables make up a mnemonic that contains the first letters of each. For example: High Winds Sometimes Wave Big Green Trees.

- **Hydroelectric Power (HEP)** is electricity generated using fast flowing water to turn the turbines. Dams are built to form reservoirs of water in high locations. The water is channeled down pipes to the power station gaining kinetic energy.
- **Wind turbines** use the energy of the wind to turn the turbines. Wind farms are collections of wind turbines.
- **Solar cells** are not used to turn turbines. Instead they produce electricity inside the cell.
- **Wave generators** use the movement of waves to generate electricity.
- **Biofuels**, or **biomass** fuels (see page 12), can be burned, or they can be used to produce methane gas or alcohol.
- **Geothermal energy** is used in power stations in places where the Earth's crust is thin and the heat is close to the surface. Geothermal power stations can use this heat.
- **Tides** change the height of the water in some areas so much that it is worth using it to generate electricity.

Advantages and disadvantages of renewable resources

AQA   P1   ✓

Source	Advantages	Disadvantages
HEP	No waste or air pollution. No fuel cost. Can generate a lot of electricity in mountainous areas.	Rainfall or snow is not constant. Building dams and flooding valleys changes the environment. Homes, farmland and natural habitats are lost forever.
Wind	No polluting waste gases. The wind is free, so the cost of electricity is low.	The wind does not always blow. Some people consider wind turbines noisy and an eyesore, and they take up a lot of space for the amount of electricity generated.
Solar cells	No air pollution. No fuel cost. No moving parts so they do not need much maintenance. They have a long life. They can be used in remote locations.	Cannot produce power at night or in bad weather.

Make sure you can explain why it is important to have a reliable supply of electricity and why this means it is best to have a mixture of ways of generating electricity and not have to rely on one method.

Source	Advantages	Disadvantages
Wave generators	No air pollution. No fuel cost. Lots of available energy.	Waves are destructive, so it has been difficult to develop the technology.
Biofuels	A way of disposing of waste. In the case of waste and manure, the pollution would be produced anyway. Carbon neutral.	Power stations need a steady supply. They produce pollution – carbon dioxide, other gases and ashes.
Geothermal	No air pollution. No fuel cost.	Earthquakes and volcanic action may damage the power station.
Tides	The tide does not depend on the weather.	The habitats of many birds and other animals may be destroyed.

PROGRESS CHECK

1. Explain how a hydroelectric power station works.
2. Choose two renewable energy sources that are suitable for use in the UK. Give your reasons for choosing them.
3. Choose two renewable energy sources that are not suitable for use in the UK. Give your reasons for deciding they are not suitable.
4. Jenny says that it would be best to generate all our electricity from wind farms and wave generators.
   a) List some advantages of wind farms and wave generators.
   b) Explain some reasons why Jenny's idea is not a good one.
5. Explain which energy source is most suitable for:
   a) Providing an electric light for a rural bus stop.
   b) Providing energy to mine and process aluminium ore in a mountainous region.

1. Rain or snow fills lakes or reservoirs high up on mountains. This water falls through fast rivers or pipes gaining kinetic energy. It is used to turn the turbines at the hydroelectric power station, which turn the generators to generate electricity.

2. Examples of suitable choices:
Wind turbines: Offshore windfarms can make use of the high winds around the coast.
Wave generators: Once suitable technology has been developed we have lots of sites where ocean waves can be exploited by wave generators.
Solar cells: Newer technology means that solar cells can produce useful amounts of energy in the UK.

3. Examples of unsuitable choices:
Hydroelectric power because there are no high mountains, so we only have very small amounts of HEP.
Tidal energy: There are no large tidal changes. The Severn estuary would be the best site, but it is an important site for wildlife.

4. a) No carbon dioxide emitted. Renewable-fuel will not run out. Some good sites round the coast.
   b) Winds and waves very variable. Often little wind when there is very cold weather. Lots of space required to generate a small amount of electricity. Lots of overhead power lines needed to connect them all to the grid. Some people consider wind turbines an eyesore and they can be noisy. It is best to have a mix of energy generation methods to avoid a complete loss of production, for example if there is a very damaging storm.

5. a) A solar panel would not need connecting to the grid and can charge batteries during the day for use at night. A light does not require much energy.
   b) This requires a lot of electricity. An HEP station could be built because the area is mountainous, and it would generate a large amount.

1.4 Electrical energy and power

LEARNING SUMMARY

After studying this section, you should be able to:

- Calculate electric power and energy.
- Use joules and kilowatt hours as units of energy.
- Calculate the cost of using electrical appliances.
- Use the equation $P = I \times V$ to calculate power, current or voltage.

Energy and power

AQA	P1	✓
OCR A	P3	✓
OCR B	P2	✓
EDEXCEL	P1, P2	✓
WJEC	P1	✓
CCEA	P2	✓

We use electrical appliances at home to transfer energy from the mains supply to:

- heating
- light
- movement and sound.

In two hours an electric lamp uses twice as much energy as it uses in one hour.

> **KEY POINT**
>
> The **power** of an electrical appliance tells us how much electrical energy it transfers in a second.
>
> Power, P is measured in **watts** (W) where 1 W = 1 J/s.

Appliances used for heating have a much higher rating than those used to produce light or sound.

Power ratings of electrical appliances.

2 kW   1 kW   800 W

The amount of energy transferred in the mains appliance depends on the power rating of the appliance and the length of time for which it is switched on.

> **KEY POINT**
>
> **Energy** transferred is worked out by:
>
> **Energy = power × time**
> $E = P \times t$
> Energy, E is measured in:
>
> - **joules** (J) when the power is in watts and the time, t, is in seconds.
> - **kilowatt hours** (kWh) when the power is in kilowatts and the time, t, is in hours.

Example: A 300 W electric pump is switched on for 1 minute. The energy used is:
E = 300 W × 60 s
E = 18 000 J

The power used by an electrical appliance is the rate at which it transfers electrical energy. A 100 W light bulb uses more electrical energy than a 60 W light bulb every second. It transfers energy at a faster rate.

Paying for electricity

AQA	P1	✓
OCR A	P3	✓
OCR B	P2	✓
EDEXCEL	P1	✓
WJEC	P1	✓
CCEA	P2	✓

The units on an electricity bill, and measured by an electricity meter, are kilowatt hours. The cost of a unit of electricity varies. The electricity bill is calculated by working out the number of units used and multiplying by the cost of a unit.

> **KEY POINT**
>
> **cost of electrical energy used = power in kW × time in hours × cost of one unit**
>
> or
>
> **cost = number of kW h used × cost of one unit**

> Remember that the kilowatt hour is a unit of energy – not power. Power is measured in watts or kilowatts.

Example: The 800 W toaster in the diagram on the previous page is used for half an hour and the cost of a unit is 12p:
1 kW = 1000 W
Cost = 0.8 kW × 0.5 hours × 12 p/kW h
Cost = 4.8 p = 5 p to nearest 1p

Power tells you how quickly energy is being used. A 3 kW fire uses the same energy in 1 hour as a 1 kW fire does in 3 hours. The energy is power × time

3kW fire: 3kW × 1h = 3kWh

1kW fire: 1kW × 3h = 3kWh

> The unit is kilowatts multiplied by hours = kWh.
>
> A common mistake is to say kilowatts per hour kW/h.

Power, current and voltage

AQA	P2	✓
OCR A	P3	✓
OCR B	P2	✓
EDEXCEL	P1, P2	✓
WJEC	P1, P3	✓
CCEA	P2	✓

The mains voltage in the UK is 230 V. Electrical power depends on the current and the voltage:

KEY POINT

Power = current × voltage

$P = I \times V$

Power is measured in watts (W), current, I, in **amps** (A) and voltage, V, in **volts** (V).

A torch with a 3.0 V battery has a current of 0.3 A. Its power is:

$P = 3.0 \times 0.3 = 0.9$ W

Calculating the current

AQA	P2	✓
OCR A	P3	✓
OCR B	P2	✓
EDEXCEL	P1	✓
WJEC	P1	✓
CCEA	P2	✓

To calculate the current in a 2 kW kettle re-arrange the equation to give:

$I = \dfrac{P}{V}$

$\dfrac{2000 \text{ W}}{230 \text{ V}} = 8.7$ A

Be careful with time when doing calculations. Change minutes to seconds if you are using energy in joules and to hours if the energy is in kilowatt hours.

PROGRESS CHECK

1. What unit is power measured in?
2. What is the power of a 230 V lamp with a current of 0.05 A?
3. Referring to the figure on page 17 how much energy in kWh does the kettle in the diagram use in six minutes?
4. Referring to the figure on page 17 how much energy in joules does the toaster in the diagram use in one minute?
5. What is the current in a 1 kW microwave oven?
6. A unit costs 12p. How much does it cost to use a 1kW microwave oven for 15 minutes?

6. units = 1kW x 0.25h = 0.25 kWh cost = 0.25 x 12p = 3p
5. 1000W/230V = 4.3 A
4. 800 W x 60s= 48 000 J
3. 2kW x 0.1 h = 0.2 kWh
2. 11.5 W
1. Watts (W)

1.5 Electricity matters

LEARNING SUMMARY

After studying this section, you should be able to:

- Describe the National Grid.
- Explain how transformers are used to reduce power loss.
- Describe how nuclear power stations work.
- Give advantages and disadvantages of nuclear fuel compared to fossil fuel.

The National Grid

AQA	P1	✓
OCR A	P3, P5	✓
OCR B	P2, P6	✓
EDEXCEL	P1	✓
WJEC	P1	✓
CCEA	P2	✓

In the UK the **National Grid** is a network that connects all the generators of electricity, like power stations, to all the users, for example homes and workplaces.

> **KEY POINT**
>
> Mains electricity is an **alternating current (a.c.)** which means that the **current** keeps changing direction. Current from batteries is **direct current (d.c.)** meaning it always flows in the same direction.

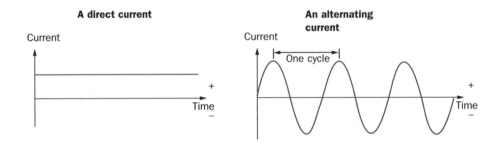

The National Grid.

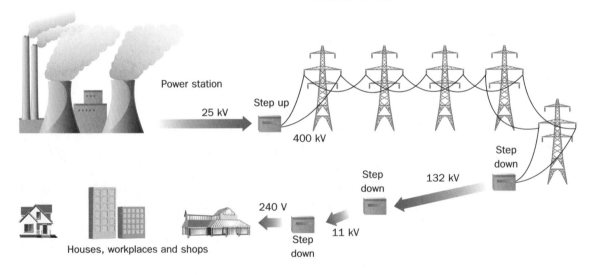

Advantages of having a National Grid are:

- Power stations can be built where the fuel reserves are, or near the sea or rivers for cooling.
- Pollution can be kept away from cities.
- Power can be diverted to where it is needed, if there is high demand or a breakdown.

For OCR A you need to be able to use all the information in this chapter to discuss energy choices in the home, national and global contexts.

A disadvantage of the National Grid is that power is wasted as heat energy in the power lines.

Transformers

AQA	P1	✓
OCR A	P3, P5	✓
OCR B	P2, P6	✓
EDEXCEL	P1	✓
WJEC	P1, P3	✓
CCEA	P2	✓

A **transformer** changes the size of an **alternating voltage**. Transformers will not work with a constant voltage. One of the reasons we have an a.c. mains supply is that the voltage is alternating and can be changed using transformers. This means it can be distributed more efficiently by using high voltages, for example 400 kV (1 kV = 1000 V).

Step-up transformers increase voltage and **step-down transformers** decrease voltage.

The voltage is stepped-up at the power station, transmitted at high voltage to reduce power losses, and stepped down at the local sub-station.

Reducing power loss in cables

AQA	P1	✓
OCR A	P3, P5	✓
OCR B	P2, P6	✓
EDEXCEL	P1	✓
WJEC	P1	✓
CCEA	P2	✓

Example: To supply 100 kW of power we can use:

High voltage and small current: $P = 100$ kW $= 1$ A $\times 100$ kV
Low voltage and large current: $P = 100$ kW $= 100$ A $\times 1$ kV

The heating effect in the cables depends on the current. By making the current as small as possible the energy wasted as heat in the cables is reduced. The current can be small if the voltage is high. There are many hundreds of miles of overhead power cables so this saves a lot of energy.

Nuclear power stations

AQA	P1	✓
OCR A	P3	✓
OCR B	P2	✓
EDEXCEL	P2	✓
WJEC	P1	✓
CCEA	P1	✓

> **KEY POINT**
>
> Nuclear power stations are like fossil fuel power stations, but instead of burning fuel they use a nuclear reaction, **nuclear fission**, to transfer energy as heat. Nuclear fuels are **uranium** and **plutonium**, which are radioactive.

Uranium is mined. Plutonium is formed in nuclear reactors. A disadvantage is that **radioactive waste** is produced that remains dangerous to living things for millions of years.

Radioactive materials emit **ionising radiation**. This is dangerous to living things (see page 58).

The nuclear option

AQA	P1	✓
OCR B	P2	✓
EDEXCEL	P2	✓
WJEC	P1	✓
CCEA	P1	✓

Advantages of nuclear power stations include:

- No carbon dioxide is formed in the nuclear reaction.
- A small amount of fuel releases a large amount of energy.
- The fuel is not expensive, so the running costs of nuclear power stations are not high.

List and learn the similarities and differences between fossil fuel (page 12) and nuclear power stations.

Know how to make a case both for and against building nuclear power stations.

Disadvantages include:

- There is the risk of an accidental emission of radioactive material while the power station is operating. This could happen due to human error, for example at Chernobyl, or to natural disasters, like earthquakes, for example at Fukushima in Japan.
- Both the power station and the fuel are targets for terrorists.
- The risks to living things from radioactive materials mean that there are high maintenance costs and high decommissioning costs. Radioactive waste must be stored safely for thousands of years.

PROGRESS CHECK

1. What is the difference between a.c. and d.c. electricity?
2. What is the National Grid?
3. Give two similarities and two differences between a nuclear power station and a coal-fired power station.
4. Why is radioactive waste dangerous?
5. What device is used inside a mobile phone charger to convert the 230 V supply to 12 V?
6. Why does the National Grid transmit electricity at 400 kV?

1. a.c. is alternating current – reverses direction d.c. is direct current – steady current.
2. The distribution network of cables and overhead power lines that links power stations with users of electricity.
3. Any two similarities, e.g. both have generators, turbines, boilers, both heat water to steam. Any two differences, e.g. nuclear: small amount of nuclear fuel (uranium/plutonium) coal: large amount of coal
Nuclear: no CO_2 but produces radioactive waste Coal: CO_2, lot of waste but not radioactive
Nuclear: nuclear reaction/fission releases a lot of energy Coal: burning/combustion of coal, less energy released.
4. It emits ionising radiation which can kill or damage living cells or cause them to turn cancerous.
5. A step-down transformer.
6. Transmitting at a high voltage means that a much smaller current can be used, for the same amount of power. This means that there is less heat lost in the cables, which saves a lot of energy, as there are hundreds of km of cables.

1.6 Particles and heat transfer

LEARNING SUMMARY

After studying this section, you should be able to:

- Calculate energy or temperatures using specific heat capacity.
- Explain the ways in which heat is transferred.
- Describe how buildings can be insulated.
- Explain and use the terms payback time and U-value.

Specific heat capacity

| AQA | P1 | ✓ |
| OCR B | P1 | ✓ |

When the temperature of an object increases it has gained energy. The amount of energy depends on:

- the temperature change, θ
- the mass of the object, m
- the specific heat capacity, c.

The **specific heat capacity** is different for different materials. It is the energy needed to increase the temperature of 1 kg of the material by 1°C and is measured in J/kg °C.

Energy = mass × specific heat capacity × temperature change
$$E = m \times c \times \theta$$

This is how to measure the specific heat capacity of a metal block:

- Measure the temperature and the mass of the block, m, at the beginning.
- Use an electric heater to raise the temperature of the metal block. Energy supplied, E = power × time.
- Measure the temperature of the block at the end of the heating time and calculate the increase in temperature θ.
- Calculate the specific heat capacity of the metal,

$$c = \frac{E}{m \times \theta}$$

Heat transfer

AQA	P1	✓
OCR B	P1	✓
EDEXCEL	P1	✓
WJEC	P1, P3	✓

Heat can be transferred by conduction, convection and radiation. It is only transferred from hotter things to cooler things. The bigger the temperature difference between a hot object and cold surroundings the faster the hot object will cool. In a hot solid, particles vibrate more. They collide with the particles next to them and set them vibrating. The kinetic energy is transferred from particle to particle. Metals are the best conductors. Solids are better than liquids. Gases are very poor conductors. They are insulators.

Conduction in a solid. Energy is transferred from molecule to molecule.

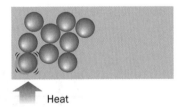

Heat

For OCR B the hot object is called the 'source' and the cold object is called the 'sink'.

In a hot fluid (gas or liquid) the particles have more kinetic energy so they move more. They spread out and the fluid becomes less dense. The hot fluid rises above the denser cold fluid forming a **convection current**.

Convection currents.

All objects emit and absorb **infrared radiation**. The higher the temperature the more they emit. When objects absorb this energy their temperature increases. Radiation will travel through a vacuum – it does not need a medium (material) to pass through.

- Dark and matt surfaces are good absorbers and emitters of infrared radiation.
- Light and shiny surfaces are poor absorbers and emitters of infrared radiation.
- Light and shiny surfaces are good reflectors of infrared radiation

The role of electrons in heat transfer

AQA	P1	✓
OCR B	P1	✓
WJEC	P3	✓

Metals are made of a lattice of positive ions with 'free' electrons that can move through the lattice. This makes them good **conductors** of heat because they have 'free' electrons to carry the energy. Because electrons are negatively charged this also makes metals good conductors of electricity.

Heat is transferred to, or from, an object at a rate that depends on:

- Its surface area and volume.
- The material it is made from.
- What the surface is like.
- The temperature difference between the object and its surroundings.

The bigger the temperature difference, the faster the object will heat up or cool down.

Insulation

AQA	P1	✓
OCR B	P1	✓
EDEXCEL	P1	✓
WJEC	P1	✓

When we insulate our homes we reduce the heat lost, we use less fuel and it costs less to heat.

Energy flow from an uninsulated house.

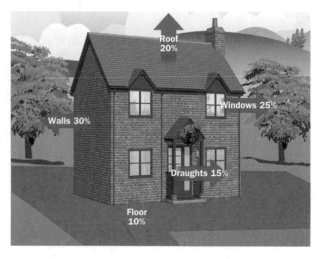

Still air is a good insulator, so materials with air trapped in them are often used.

- In cavity walls the air gap between the walls stops conduction.
- Cavity wall insulation works by filling the cavity with foam or mineral wool.
- Loft insulation uses layers of fibreglass or mineral wool.
- Reflective foil on walls reflects infrared radiation.
- Draught-proofing stops hot air leaving and cold air entering the house.

> Convection can occur in the cavities in cavity walls – cavity wall insulation stops this.

All these improvements cost money to buy and install, but they save money on fuel costs. You can work out the **payback time**, which is the time it takes before the money spent on improvements is balanced by the fuel savings, and you begin to save money.

KEY POINT

$$\text{payback time (in years)} = \frac{\text{cost of insulation}}{\text{cost of fuel saved each year}}$$

If the price of the fuel increases, the payback time will be less.

U-values

AQA **P1** ✓

The materials used in constructing a building, like glass, brick, wood and concrete, are given a **U-value** to indicate how good they are at insulating. The lower the U-value the better the insulator.

PROGRESS CHECK

1. How much energy is needed to raise the temperature of 1 kg of water by 2°C? (water, c = 4200 J/kg°C)
2. Draw and label convection currents in a beaker of water heated by a Bunsen burner.
3. A new boiler costs £2000. It saves £100 on fuel costs each year. What is the payback time?
4. 4400 J of energy raises the temperature of 0.5 kg of aluminium from 15°C to 25°C. Calculate the specific heat capacity of aluminium.
5. How does cavity wall insulation work?

5. Traps the air in an insulating foam so that conduction and convection are stopped.
4. 4400 J = 0.5kg × c × 10 °C c = 4400 J ÷ (0.5 kg × 10 °C) = 880 J/kg°C
3. £2000 ÷ £100 per year = 20 years
2. Diagram similar to convection current in house shown on page 24. Hot water rises, cold water sinks.
1. E = 1 kg × 4200 J/kg°C × 2°C = 8400 J

Sample GCSE questions

1 A coal-fired power station generates electricity.

(a) Complete the boxes to describe the processes in a coal fired power station. **[5]**

Furnace	Boiler	Turbine	Generator	Transformer
burns the coal	Heats water	turned by steam and rotates the generator	generates electricity	changes the voltage

(b) A coal fired power station uses 1500 MW of power from coal to produce 555 MW.

Calculate the efficiency of the power station. **[1]**

$$\text{efficiency} = \frac{555}{1500} \times 100\% = 37\%$$

← Always show your working

(c) Energy is wasted in the power station and also in the transmission lines. Complete this Sankey diagram to show

- where the energy is wasted in the power station each second
- how much energy reaches the customers each second. **[2]**

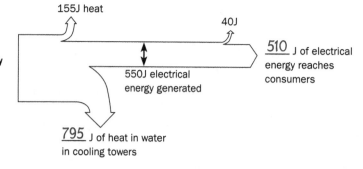

155J heat

40J

1500J energy in coal

550J electrical energy generated

510 J of electrical energy reaches consumers

795 J of heat in water in cooling towers

(d) An 3kW electric fire is switched on for 4 hours every evening for 2 weeks. A unit of electricity costs 12p.

Calculate the bill for using the fire. **[3]**

3kW x (4 hours x 7 x 2) = 168 kWh
cost = 168 kWh x 12p per kWh = £20.16

(e) The National Grid has hundreds of miles of power lines. Explain how the energy lost in heating these overhead power lines is reduced. *The quality of your written communication will be assessed in this answer.* **[6]**

Transformers (1 point) are used to change the voltage (1 point). Step-up transformers increase voltage (1 point) at the power station (1 point) to very high voltages (1 point). The power is transmitted at high voltage (1 point). Step-down transformers decrease voltage at the substation (1 point).

← Marks will be awarded depending on the number of relevant points included in the answer and the spelling, punctuation and grammar. In this question there are 12 relevant points so 10 or 12 with good spelling punctuation and grammar will gain full marks.

Sample GCSE questions

Power = voltage x current (1 point), so using high voltage means a low current (1 point) can be used for the same power (1 point). Low current means less energy is wasted (1 point) as heat in the power lines (1 point).

[Total = 17]

2 A public enquiry has been set up to decide whether to build a nuclear power station or a coal-fired power station.

The coal-fired power station would release a gas that most scientists think may contribute to climate change.

(a) What is the name of the gas? **[1]**

carbon dioxide

(b) Explain how the gas may contribute to climate change. **[2]**

absorbs infrared radiation and warms the atmosphere

(c) Choose A, B, C or D. Tick the correct answer.
The fuel that is used in some nuclear reactors is:
A Caesium   **C** Uranium ✓
B Radium   **D** Vanadium **[1]**

(d) Some local people are consulted about their views.

Andy

In an earthquake or an accident there could be a release of radioactive material.

Bella

Climate change is a much more serious threat to life on Earth than ionising radiation from nuclear waste.

Chloe

A nuclear power station can produce more electricity than a coal-fired power station.

Dan

Radioactive waste has to be kept safe for hundreds of years.

(i) Who is talking about a risk of nuclear power?   Andy and Dan

(ii) Who is talking about a risk of coal-fired power?   Bella

(iii) Who is talking about a benefit of nuclear power? Chloe   **[3]**

(e) To decide which power station to build a 'Life Cycle Assessment' is made. State three factors considered in a 'Life Cycle Assessment'. **[3]**

It takes into account all the energy costs of building (1 mark), running (1 mark), and decommissioning (1 mark) the power station.

[Total = 10]

Exam practice questions

1 Which of the following is a renewable energy source? [1]

- **A** Coal ☐
- **B** Gas ☐
- **C** Nuclear ☐
- **D** Wind ☐

2 Which of the following energy sources is associated with CO_2 emissions? [1]

- **A** Gas ☐
- **B** Nuclear ☐
- **C** Solar ☐
- **D** Wind ☐

3 Explain why the following statements are true.

(a) A carpet feels warmer to bare feet than a stone floor.

....

(b) Two thin blankets are usually warmer than one thick one.

....

(c) You should crawl on the floor to escape a smoke filled room.

....

(d) The element of an electric kettle is in the bottom of the kettle.

....

(e) The cooling pipes on the back of a refrigerator are painted black.

....

[Total = 5]

4 **(a)** Complete this diagram of a coal-fired power station.

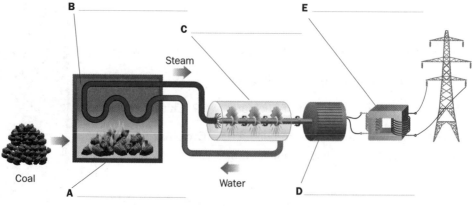

[5]

Exam practice questions

(b)   What fuel is used in a nuclear power station?   **[1]**

....

(c)   Describe how a nuclear power station works differently to a coal-fired power station.   **[2]**

....

....

(d)   Explain one advantage and one disadvantage of a nuclear power station compared to a coal-fired power station.   **[5]**

....

....

[Total = 13]

5   This table shows some electrical power ratings for wind generators.

Model	Maximum power rating of one turbine (kW)	Wind turbine recommended for:
A	0.025	Battery charger
B	0.5	Electricity for a caravan
C	5	Domestic electricity for a house
D	25	Electricity for a school
E	500	One for a wind farm
F	1500	One for an offshore wind farm

(a)   What will affect the amount of power that a wind turbine can generate?   **[1]**

....

(b)   What model is recommended for a house?   **[1]**

....

(c)   A house has an average power use of 0.5 kW. Describe how the actual power used might be

    (i)   higher than average   **[1]**

....

    (ii)   lower than average.   **[1]**

....

    (ii)   Suggest two reasons why model B is not recommended for this house.   **[2]**

....

....

Exam practice questions

(e) Wind turbines are often used to recharge banks of batteries. Why is this a good idea? **[1]**

....

....

(f) A power station generates 750 MW of electrical power. How many wind turbines would be needed to replace it with

(i) an offshore wind farm? **[1]**

....

(ii) an onshore wind farm? **[1]**

....

(g) What is meant by 'the payback time?' **[1]**

....

....

[Total = 10]

6 The electrical power used by an appliance is **[1]**

A the efficiency per second. ☐
B the electric current per second. ☐
C the energy per second. ☐
D the voltage per second. ☐

7 This diagram shows an experiment where a magnet is inserted in a coil.

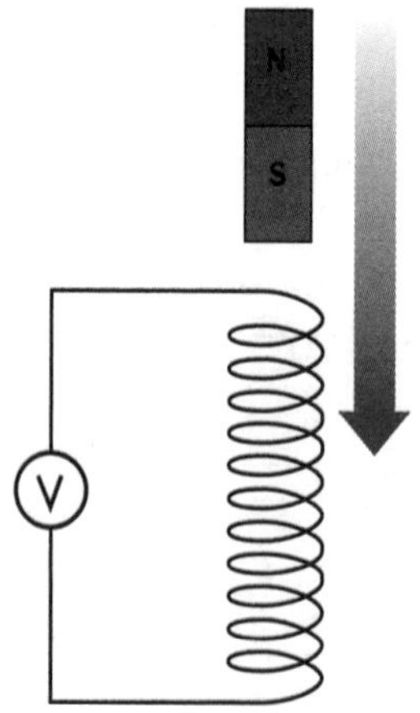

The voltmeter registers a maximum voltage of about + 0.5 mV as the magnet is moved into the coil.

Describe how the voltmeter reading changes when the experiment is repeated and:

(a) the magnet is inserted more quickly into the coil. **[1]**

....

Exam practice questions

(b) the magnet remains stationary inside the coil. **[1]**

....

(c) the magnet is removed from the coil. **[1]**

....

(d) the magnet is reversed so that the south pole enters the coil first. **[1]**

....

[Total = 4]

8 A 100 W light bulb uses 100 W of electrical power to produce 8 W of light energy.

(a) Calculate the efficiency of the light bulb. **[2]**

....

....

(b) Explain what happens to the rest of the electrical energy supplied to the light bulb. **[1]**

....

(c) An energy saving lamp has an efficiency of 40% and gives the same output power of 8 W. What electrical power does it use? **[2]**

....

....

(d) A student uses a desk lamp for 3 hours a day. How many kilowatt hours of electricity does the lamp use in a week if it is fitted with a 100 W light bulb? **[2]**

....

....

....

[Total = 7]

9 A public enquiry has been set up to decide whether to build a new gas-fired power station or a nuclear power station. Describe the advantages and disadvantages of a nuclear power station compared to a gas-fired power station. *The quality of written communication will be assessed in your answer*.

....

....

....

....

....

[Total = 6]

2 Waves

The following topics are covered in this chapter:

- **Describing waves**
- **Wave behaviour**
- **Seismic waves and the Earth**

2.1 Describing waves

LEARNING SUMMARY

After studying this section, you should be able to:

- Explain the difference between longitudinal and transverse waves.
- Give examples of longitudinal and transverse waves.
- Describe waves, using words like frequency, wavelength and amplitude.
- Calculate the speed of waves, their frequency or wavelength.

Transverse and longitudinal waves

AQA	P1	✓
OCR A	P1	✓
OCR B	P1, P4,	✓
	P5	✓
EDEXCEL	P1	✓
WJEC	P1, P3	✓
CCEA	P2	

A **wave** is a **vibration** or disturbance transmitted through a material (a medium) or through space. Waves transfer energy and information from one place to another, but they do not transfer material.

A **transverse wave** has vibrations at right angles (perpendicular) to the direction of travel. The wave has **crests** and **troughs**. Examples include water waves, waves on strings or rope, light and other electromagnetic waves.

A transverse wave.

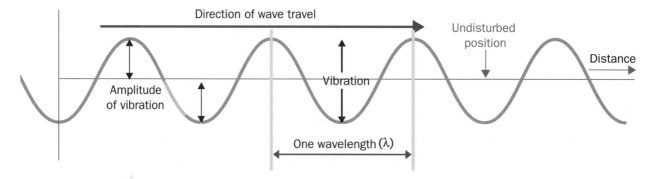

A **longitudinal wave** has vibrations parallel to the direction of wave travel. It has **compressions** and between these are stretched parts called **rarefactions**. Examples include sound waves, **ultrasound** waves, which are sound waves with frequency greater than 20 kHz, and waves along a spring. Infrasound waves are waves with a frequency lower than 20 Hz.

Ultrasound is used for sonar, animal communication and fetal scanning. Infrasound is used for animal communication and detecting disturbances like animal movement and meteors.

A longitudinal wave.

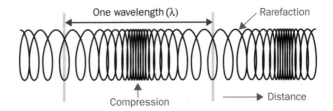

One wavelength (λ)

Rarefaction

Compression

Distance

Wave travel and properties

AQA	P1	✓
OCR A	P1	✓
OCR B	P1	✓
EDEXCEL	P1	✓
WJEC	P1	✓
CCEA	P2	✓

Longitudinal waves travel through solids, liquids and gases, but they cannot travel through a vacuum.

Transverse **electromagnetic waves** can travel through a vacuum. They are oscillations of a magnetic and electric field.

There are two types of **seismic waves** (see page 38) called **P-waves** and **S-waves**:

- **P-waves** are longitudinal waves that travel through solid or liquid rock. They travel faster than S-waves.
- **S-waves** are transverse waves that can only travel through solid materials in the Earth

> A common mistake is to mark the amplitude from the top of a peak to the bottom of a trough – this is twice the amplitude.

Amplitude is the maximum displacement (change in position) from the undisturbed position.

Wavelength (λ) is the distance in metres from any point on the wave to where it repeats.

Frequency is the number of waves that pass in one second. This depends on how fast the source of the waves is vibrating. The frequency is usually expressed in hertz (Hz) where one Hz is one cycle (wave) per second.

Example: Four waves pass a point in one second. The frequency = 4 Hz.

Example: A wave takes two seconds to pass a point. The frequency = 0.5 Hz.

> Use words like the ones below to explain what you mean, so the examiner can award you the marks.
> - **Oscillation** for side to side, up and down, or back and forth movements.
> - **Perpendicular** for at right angles.
> - **Parallel** for in the same direction.

Wave speed depends on the medium that the wave is travelling through.

distance wave travels = wave speed × time ($d = v \times t$)

Example: A water wave travels at 5 cm/s.
In 4 s it travels d = 5 cm/s × 4s = 20 cm.

> **KEY POINT**
>
> The **wave equation** relates the wavelength and frequency to the wave speed. For all waves:
>
> **wave speed (v) = frequency (f) × wavelength (λ)** ($v = f\lambda$, where f is in Hz, λ is in m, v is in m/s).

Example: Water waves with wavelength λ = 10 cm and frequency *f* = 4Hz have a speed *v* = 10 cm × 4 Hz = 40 cm/s.

Waves with different frequency and wavelength.

A wave on a rope   A higher frequency wave

Using the wave equation

AQA	P1	✓
OCR A	P1	✓
OCR B	P1	✓
EDEXCEL	P1	✓
WJEC	P1	✓
CCEA	P2	✓

The wavelength of light is very small and the wave speed and frequency are very high.

Example: Light travels at 300 000 000 m/s = 3 × 10⁸ m/s.

Wave speed *v* = 3×10^8 m/s.

Green light has wavelength, λ = 520 nm = 5.2×10^{-7} m

(1 nm is 1 nanometre = 1×10^{-9}m).

To calculate the frequency, *f*, of the light waves, re-arrange the wave equation:

$$f = \frac{v}{\lambda}$$

$$f = \frac{3 \times 10^8 \text{ m/s}}{5.2 \times 10^{-7}\text{m}} = 5.8 \times 10^{14} \text{ Hz}$$

Sound waves

AQA	P1	✓
OCR B	P4	✓
EDEXCEL	P1	✓
CCEA	P2	✓

For sound and ultrasound waves:

- The **pitch** of the sound is the frequency of the vibrations.
- The **loudness** of the sound depends on the amplitude of the vibrations.

Sound travels much faster in solids than in liquids and faster in liquids than in gases.

> **PROGRESS CHECK**
>
> 1. Draw a transverse wave.
> 2. Mark on the wave the amplitude and the wavelength.
> 3. How is a longitudinal wave different to a transverse wave?
> 4. A sound wave has frequency 256 Hz and wavelength 1.30 m. Calculate the speed of the sound.
> 5. Water waves pass a marker at a rate of one every four seconds. What is their frequency?
> 6. What is the frequency of radio waves with a wavelength of 3 m?

2.2 Wave behaviour

LEARNING SUMMARY

After studying this section, you should be able to:

● Describe reflection and the images formed by a plane mirror.
● Describe and explain refraction.
● Describe and explain diffraction.
● Draw ray diagrams and wave diagrams.

Reflection

AQA	P1	✓
EDEXCEL	P1	✓
WJEC	P3	✓
CCEA	P2	✓

All waves can be reflected, but this does not prove they are waves because particles also show these effects. According to the **law of reflection**, the angle of incidence equals the angle of reflection. This can be shown on a wave diagram or a ray diagram.

Reflection.

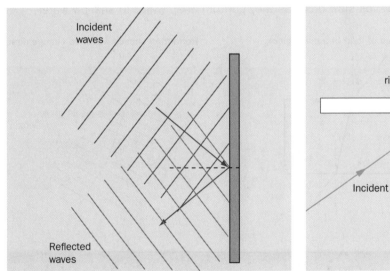
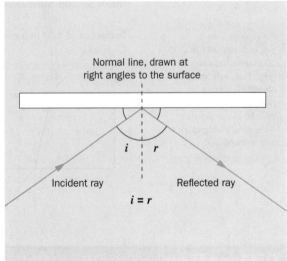

Echoes are reflections of sound waves from hard surfaces.

The image in a mirror

| AQA | P1 | ✓ |
| CCEA | P2 | ✓ |

The **image** in a plane (flat) mirror is the same distance behind the mirror as the object is in front. It is a virtual image. The image is also upright and laterally inverted (left becomes right).

Remember that the angles of incidence, reflection and refraction are all measured to the normal, not to the surface. Draw the normal at right angles to the surface.

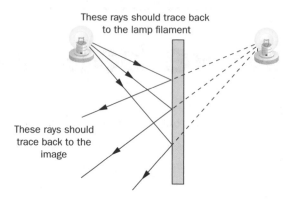

The image in a plane mirror.

These rays should trace back to the lamp filament

These rays should trace back to the image

Refraction

AQA	P1	✓
OCR A	P7	✓
OCR B	P1, P5	✓
EDEXCEL	P1, P3	✓
WJEC	P3	✓
CCEA	P2	✓

When waves enter a different **medium** they change **speed**. The wavelength changes, but the frequency stays the same. They change direction unless they are travelling along the normal to the boundary. This is called **refraction**.

As light waves enter a **denser medium** they slow down and bend towards the normal. As light waves enter a **less dense medium** they speed up and bend away from the normal.

Water waves slow down as they go from deep water to shallow water.

Make sure that you understand the difference between reflection, refraction and diffraction. Students often muddle these words – especially refraction and diffraction.

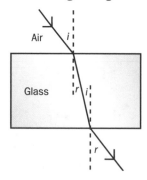

Refraction of light in a glass block.

Air

Glass

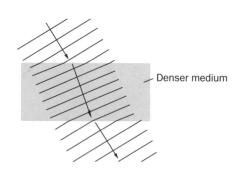

Refraction causes waves to change direction

Denser medium

Diffraction

AQA	P1	✓
OCR A	P7	✓
OCR B	P1, P5	✓

Diffraction is the spreading out of a wave when it passes through a gap. The effect is most noticeable when the gap is the same size as the wavelength. Particles cannot be diffracted, so diffraction is good evidence for the existence of a wave.

Diffraction.

a) 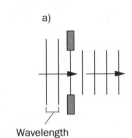   b)   c)

Wavelength

It is important to compare the size of the gap with the wavelength when you are explaining whether diffraction will occur and when the effect will be biggest.

Diffraction effects

| AQA | P1 | ✓ |
| OCR B | P5 | ✓ |

The wavelength of sound waves is about a metre, so when sound waves are diffracted through doorways people can hear round corners.

Water waves are sometimes diffracted by harbour entrances.

The wavelength of light is about half of a millionth of a metre – so small that diffraction effects are only noticeable when light waves pass through gaps of about a hair's width. Larger gaps have shadows with sharp edges. This is one reason why it took so long for scientists to realise that light can behave as a wave.

PROGRESS CHECK

1. If the angle of incidence is 30°, what is the angle of reflection?
2. A ray goes from air to glass with an angle of incidence of 30°. Is the angle of refraction bigger, smaller or the same?
3. Draw a ray diagram to show light at an angle to the normal crossing a boundary from a) air to glass and b) from water to air.
4. Draw a ray diagram to show how two mirrors can be used to see round a corner.
5. Sketch a diagram of:
   a) Water waves with wavelength 1 cm passing through a 1 cm gap.
   b) Light waves with wavelength 5×10^{-7} m passing through a 1 mm gap.
   c) The same light waves passing through a 1 micrometre gap $(1 \times 10^{-6}$ m).

5. a) like diagram c) top of page 37
   b) like diagram a) top of page 37
   c) like diagram b) top of page 37

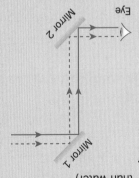

4.

3. As diagram on page 36 a) top boundary b) bottom boundary (air is less dense than water)

2. smaller

1. 30°

2.3 Seismic waves and the Earth

LEARNING SUMMARY

After studying this section, you should be able to:

- Explain the difference between S-waves and P-waves.
- Interpret a seismograph.
- Explain how S-waves and P-waves can give scientists information about the structure of the Earth.
- Describe the structure of the Earth and plate tectonics.

Seismic waves

OCR A	P1	✓
OCR B	P1	✓
EDEXCEL	P1	✓
WJEC	P3	✓

Seismic waves are caused by earthquakes or large explosions, such as quarrying operations or atomic bomb tests. P-waves and S-waves both travel *through* the Earth and are detected at monitoring stations all around the Earth by instruments called **seismometers**. They record the waves arriving on a **seismograph**.

P-waves travel faster through the Earth than other seismic waves, so they are the first to be detected after an earthquake.

S-waves are detected after P-waves.

The time delay between the arrival of the P-wave and S-waves is greater at stations further from the earthquake, so the distance to the earthquake centre can be calculated. Using information from several stations the position of the earthquake is worked out.

For Edexcel you need to explain that it is difficult for scientists to predict earthquakes because there is little advance warning and the time scales are very long – 'sometime during the next 50 years'.

A seismograph

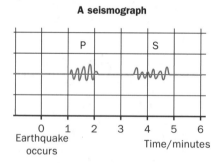

The structure of the Earth's crust can be investigated by setting up monitoring equipment at different points and setting off a controlled explosion. The waves are then recorded arriving at the monitoring points. Analysing this data gives information about the rock structure. The speed of the waves is different in different materials and waves will be reflected and refracted at boundaries between different types of rock.

Information from waves

OCR A	P1	✓
OCR B	P1	✓
EDEXCEL	P1	✓
WJEC	P3	✓

Data from all the monitoring stations after earthquakes shows a **shadow zone** on the opposite side of the Earth to the earthquake where no S-waves are detected. This can be explained by the Earth having a liquid core, because S-waves don't travel through liquids.

The diagram below shows the paths of P-waves and S-waves through the Earth following an earthquake. The waves are reflected and refracted at the boundaries.

> Think of P-waves as primary waves and S-waves as secondary waves. After an earthquake, P-waves arrive first and S-waves arrive second. P-waves have a faster speed.

P and S waves travel through the Earth.

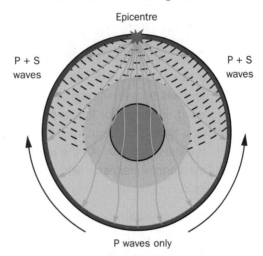

The rock cycle

OCR A	P1	✓
WJEC	P3	✓

The Earth must be older than the oldest rocks, which scientists believe are about four thousand million years old. Rock processes seen today can explain changes that happened in the past. Rocks are worn down by erosion. Sediments are deposited in layers by rivers and seas and form new rocks. All the continents would be worn flat if new mountains were not being formed.

The structure of the Earth

OCR A	P1	✓
EDEXCEL	P1	✓
CCEA	P2	✓

The Earth is made up of a **core**, a **mantle** and a **crust**.

The Earth's structure.

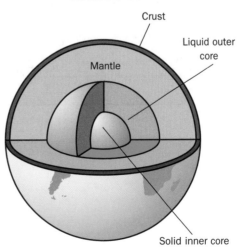

> For CCEA you need to know that the crust and upper part of the mantle is called the lithosphere.

The theory of continental drift was first suggested by Alfred Wegener in 1912, based on the way the continents appeared to fit together like a jigsaw, and matching fossils and rock layers on different continents. Due to lack of evidence, it was not until 1967 that the theory of plate tectonics was accepted.

The Earth's crust consists of a number of moving sections called **tectonic plates**. The mantle behaves like a very thick liquid. The plates move because of convection currents in the mantle.

> **Be able to explain that Wegener's theory was not believed because:**
>
> - He was not a geologist.
> - There was no explanation of how the continents move – no mechanism.
> - There was no way of measuring any movement – no evidence.

Plate tectonics

OCR A   P1   ✓
EDEXCEL P1   ✓

Magma flows out of the mid-ocean ridges forming new rock, so the sea floor spreads by a few centimetres a year and this causes the continents to move apart. Evidence for this is found in the new rocks at the mid-ocean ridge. The new rocks are rich in iron. Every few thousand years the Earth's magnetic field reverses direction. As the rocks solidify they are magnetised in the direction of the Earth's magnetic field, so the rocks contain a magnetic record of the Earth's field.

At boundaries, where two plates collide, rocks are pushed up forming new mountain ranges or when plates slide past each other this can sometimes cause earthquakes. At boundaries where magma comes to the surface there are volcanoes. This is why earthquakes, mountains and volcanoes are generally found along the edges of the tectonic plates.

> **PROGRESS CHECK**
>
> 1. Which arrives first after an earthquake, P-waves or S-waves?
> 2. What is a seismometer?
> 3. What does the theory of plate tectonics suggest about the surface of the Earth?
> 4. How can the distance to an earthquake be worked out from seismometer readings?
> 5. Are **a)** P-waves and **b)** S-waves transverse or longitudinal?
> 6. How do scientists know there is a liquid part of the Earth's core?
>
> 6. Because there is an S-wave shadow zone, S-waves cannot travel through liquids.
> 5. a) P waves are longitudinal b) S-waves are transverse.
> 4. The time delay between P-waves and S-waves arriving depends on how far away the earthquake is.
> 3. It's made up of a number of plates floating on the mantle.
> 2. An instrument for detecting seismic waves from e.g. earthquakes.
> 1. P-waves

Sample GCSE questions

1   This diagram shows a water wave travelling across the surface of a pond.

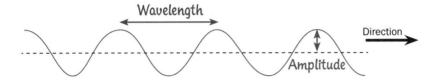

(a)   What is the name of this type of wave?   **[1]**

transverse

(b)   Mark on the wave **(i)** the amplitude **(ii)** the wavelength   **[2]**

(c)   Two wave crests pass a post in the water every second.
What is the frequency of the wave?   **[1]**

2 Hz

(d)   The wavelength of the wave is 30 cm.
Calculate the speed of the waves   **[2]**

$v = 2$ Hz x 30 cm
$v = 60$ cm/s
Wave speed $= 60$ cm/s   **[Total = 6]**

See diagram. The amplitude can be from the centre of the wave to the top of a crest or to the bottom of a trough. A wavelength is one complete wave. The starting point can be at any point on the wave, so choose an easy one, for example crest to crest.

Don't forget the units, there are no marks for '2'

Show your working. There is one mark for '2 x 30' even if your final answer is incorrectly calculated.

2   This chart is a seismograph recorded some distance away from an earthquake. It shows the arrival of P-waves and S-waves.

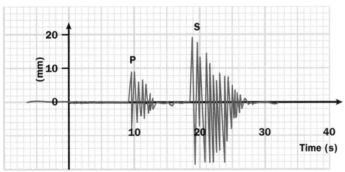

(a)   Why do the P-waves arrive before the S waves?   **[1]**

The P-waves travel faster

(b)   Describe the difference in amplitude of the waves shown on the seismograph.   **[1]**

The S-waves have larger amplitude

(c)   Give one other major difference between P-waves and S-waves.   **[1]**

P are longitudinal, S are transverse

(d)   What is the time delay between the arrival of the P-waves and the arrival of the S-waves?   **[1]**

12 s

[Total = 4]

Or P travel through liquids S do not. Remember that it is not enough to say just 'P waves travel through liquids' or 'P waves are longitudinal'.

Not 12 but 12 s or 12 seconds (accept any answers between 10–12 s)

Exam practice questions

1 Here are some statements about waves. Write **T** for **True** statements and **F** for **False** statements.

(a) Longitudinal waves have oscillations in the direction of travel. ....

(b) Transverse waves have oscillations opposite to the direction of travel. ....

(c) There is a change of wave speed during refraction. ....

(d) The frequency does not change when a wave is refracted. ....

(e) Diffraction occurs through gaps much smaller than the wavelength of the wave. ....

(f) Wave speed is equal to the frequency x the wavelength. ....

(g) Frequency is measured in Hertz per second. ....

[Total = 7]

2 These waves are travelling across a ripple tank. The wave generator moves up and down with a frequency of 4 Hz.

Wave generator | ⊢——— 45cm ———⊣

(a) What is the wavelength of the waves?

....

....

Wavelength = ... cm **[2]**

(b) Calculate the speed of the waves. Write down the equation you use and show how you calculated your answer.

....

....

Wave speed = ... cm/s **[2]**

(c) What will happen to the speed of the waves when the depth of the water is increased? **[1]**

....

[Total = 5]

3 Seismic surveying involves setting off an explosion and recording the seismic waves arriving at different points on the Earth's surface.

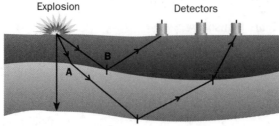

Explosion   Detectors

Exam practice questions

What is happening to the seismic waves:

(a) at point A? [1]

....

(b) at point B? [1]

....

(c) Explain how geologists use seismic surveying to get information about rock layers.
The quality of your written communication will be assessed in this answer. [6]

....

....

....

[Total = 8]

4 This diagram shows the water waves with wavelength 10 m entering a harbour.

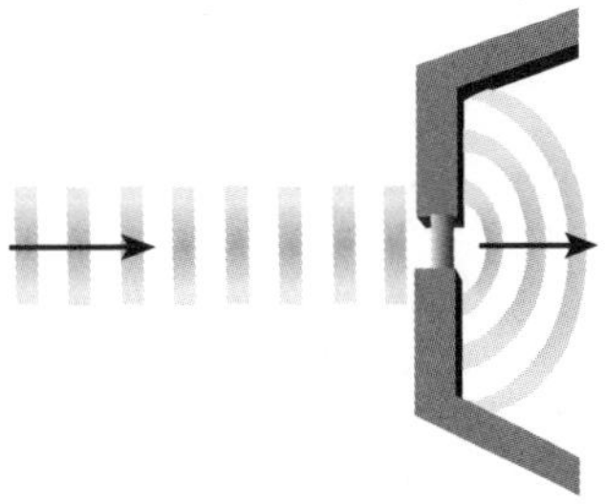

Wave direction →

(a) What is this effect called? [1]

....

(b) On another day there are smaller waves with wavelength 2 m. Describe how the waves in the harbour will be different. A diagram may help you. *The quality of your written communication will be assessed in this answer.* [6]

....

....

[Total = 7]

5 Sound waves travel at 330 m/s in air.

(a) What type of wave is a sound wave? [1]

....

(b) A musical note has frequency 256 Hz. Calculate the wavelength of the sound waves in air. [3]

....

....

[Total = 4]

Exam practice questions

6 This diagram shows the structure of the Earth.

Key	
A	
B	
C	
D	

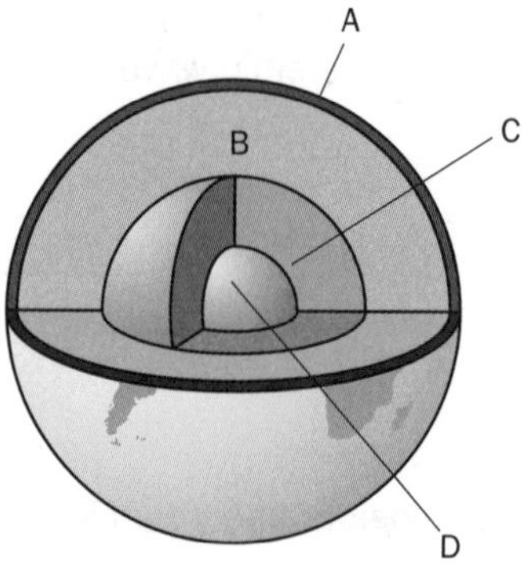

(a) Complete the key for the diagram. [4]

(b) The surface of the Earth is made up of a number of moving plates.

What are the plates called? [1]

....

(c) Explain why the plates move. *The quality of your written communication will be assessed in this answer.* [6]

....

....

....

[Total = 11]

7 Look at this seismograph trace.

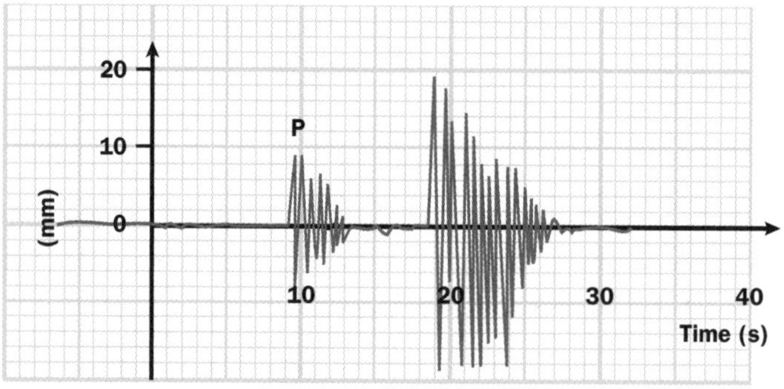

Explain how the arrival of P-waves can be used to save lives in an earthquake. *The quality of your written communication will be assessed in this answer.*

....

....

....

....

....

[Total = 6]

3 Electromagnetic waves

The following topics are covered in this chapter:

- **The electromagnetic spectrum**
- **Light, radio waves and microwaves**
- **Wireless communication**
- **Infrared**
- **The ionising radiations**
- **The atmosphere**

3.1 The electromagnetic spectrum

LEARNING SUMMARY

After studying this section, you should be able to:

- List the types of radiation in the electromagnetic spectrum in order of wavelength, frequency and energy.
- Recognise typical wavelengths for each type of radiation.
- Understand and use the term intensity.
- Calculate energy or intensity of radiation.

The electromagnetic spectrum

AQA	P1, P3 ✓
OCR A	P2 ✓
OCR B	P1, P5 ✓
EDEXCEL	P1, P3 ✓
WJEC	P1 ✓
CCEA	P2 ✓

The spectrum of electromagnetic waves is continuous from the longest wavelengths (**radio waves**) through to the shortest wavelengths (**gamma rays.**)

The spectrum goes from radio waves to gamma rays. To remember what comes in between:

- Remember the colours of visible light using the mnemonic ROY G BIV.
- Next remember infrared is under red and ultraviolet is above violet in energy.
- Then remember waves (radio and micro) are below infrared and rays (X and gamma) are above ultraviolet.

The electromagnetic spectrum

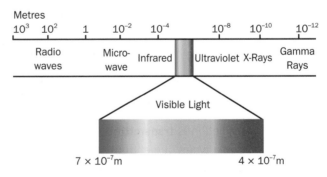

All electromagnetic waves are transverse waves that can travel through a vacuum. They all travel through empty space at a speed of 300 000 000 m/s. Radio waves have the longest wavelength, lowest frequency and lowest energy. Gamma rays have the shortest wavelength, highest frequency and highest energy. The wavelengths are related to the frequencies using the wave equation (see page 34).

Wavelength ranges

AQA	P1, P3	✓
OCR A	P2	✓
OCR B	P1	✓
EDEXCEL	P1, P3	✓
WJEC	P1	✓
CCEA	P2	✓

Remember these typical wavelengths:

- 1 km = radio
- 1 cm = microwave
- 0.1 mm = infrared
- 10^{-10} m = X-ray

The wavelength of visible light is about half a thousandth of a millimetre, which is so small that it is not obvious that it is a wave. Sometimes light behaves as a stream of particles.

There are not fixed boundaries between the different types of electromagnetic waves. The diagram on page 45 shows the typical values of wavelength for each type of wave. Gamma rays are always shown as shorter wavelength than X-rays, but the ranges overlap. The real difference is that gamma rays come from radioactive materials and X-rays are produced in an X-ray tube.

Energy and intensity

OCR A	P2	✓
EDEXCEL	P3	✓

Radiation spreads out from the source.

spotlight

A1

KEY POINT

When electromagnetic radiation strikes a surface, the **intensity** of the radiation is the amount of energy arriving at a square metre of the surface each second.

Radiation is emitted from a source and travels towards a destination. On this journey the radiation spreads out, so the further away a detector is from the source the less energy is detected. The intensity of the radiation can be increased by moving closer to the source.

Intensity can also be increased by increasing the radiation from the source.

larger area less intensity

A2

The identical spotlights on the same area double the intensity.

Spotlight   Spotlight

Double intensity

A

Radiation is **transmitted** from the source to the detector. Some materials **absorb** some types of electromagnetic radiation. The further the radiation travels through an absorbent material the lower its intensity will be when it reaches the end of its journey. More of the radiation energy is absorbed by the material and it heats up.

On some journeys, electromagnetic radiation is **reflected** at a boundary between two different materials.

When the energy of the radiation is absorbed by the detector it may:

- Have a heating effect.
- Produce small electric currents in aerials (if microwaves or radio waves).
- Make chemical reactions more likely, for example light causes photosynthesis in plants.
- Ionise atoms (if ultraviolet, X-ray or gamma).

Comparing intensities

| OCR A | P2 | ✓ |
| EDEXCEL | P3 | ✓ |

To compare the **intensity** of radiation we measure the amount of energy falling on one square metre of the surface in each second. The energy absorbed by the surface or detector can be calculated by:

Energy (J) = intensity (W/m^2) × time (s)

Electromagnetic radiation sometimes behaves as waves, and sometimes as packets of energy called **photons**. A gamma ray photon has the most energy. The energy of the photons increases with the frequency of the radiation. The energy arriving at a surface is the number of photons multiplied by the energy of the photon.

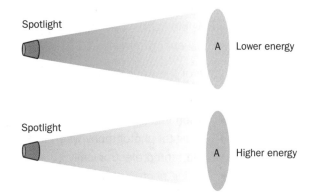

Blue light has more energy than red light.

PROGRESS CHECK

1. State two properties that all electromagnetic waves have in common.
2. State one difference between infrared and ultraviolet rays.
3. An electromagnetic wave has a wavelength of 10^{-6} m (0.001 mm). What type of electromagnetic radiation is it?
4. Which radiations have an ionising effect?
5. Microwaves from Source A have a frequency of 3 GHz and from Source B have a frequency of 30 GHz.
   a) Which microwaves have the most energy?
   b) Calculate the wavelength of the microwaves from A and B.

1. Two from: transverse; travel at 3×10^8 m/s in a vacuum; can travel through a vacuum; are oscillations of a magnetic and electric field.
2. Infrared have lower frequency or longer wavelength, or lower energy than ultraviolet. Or ultraviolet are ionising, infrared are not.
3. Infrared
4. Ultraviolet, X-rays, gamma rays
5. a) B have the most energy (higher frequency so higher energy)
   b) A = $3 \times 10^8 \div 3 \times 10^9 = 0.1$ m or 10 cm; B = $3 \times 10^8 \div 30 \times 10^9 = 0.01$ m or 1 cm

3.2 Light, radio waves and microwaves

After studying this section, you should be able to:

- Explain some properties and uses of visible light.
- Explain how the eye is similar to a camera.
- Explain some properties of radio waves and microwaves.
- Describe differences between radio waves and microwaves.
- Explain uses of microwaves and radio waves.

Visible light

AQA	P1	✓
OCR A	P2	✓
OCR B	P1	✓
EDEXCEL	P1	✓
CCEA	P2	✓

When electromagnetic radiation from the Sun arrives at the Earth's atmosphere, some of it is reflected, some is transmitted to the Earth's surface and some is absorbed by the atmosphere. The types of electromagnetic radiation that arrive at the Earth's surface are:

- high energy infrared
- visible light
- low energy ultraviolet.

Some of the ways visible light is used include:

- When images are formed on the retina at the back of the eye.
- When images are formed on the light sensitive film in a camera.
- In digital photography when images are formed on the light sensitive screen and stored electronically.
- During photosynthesis when plants use light as an energy source.

Making images

| EDEXCEL | P1 | ✓ |
| CCEA | P2 | ✓ |

Light travels in straight lines so always draw rays of light with a ruler.

How an image is formed

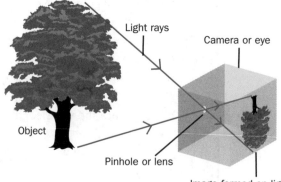

Light rays

Camera or eye

Object

Pinhole or lens

Image formed on light sensitive surface

The eye is similar to a camera. In both, light rays pass through a small hole and produce a small upside down image. A lens can be used to gather and focus more light. The image is formed on a light sensitive surface in the following ways:

- On photographic film a chemical reaction occurs which changes the colour of the film.

> The more bytes the higher the quality of the image or sound.

- On the cells of the retina a chemical reaction causes electric signals to travel along the **optic nerve** to the brain.
- On the screen inside a digital camera electronic signals are sent to the memory. The amount of information to store the picture is measured in **bytes**.

Radio waves and microwaves

AQA	P1	✓
OCR A	P2	✓
OCR B	P1	✓
EDEXCEL	P1	✓
CCEA	P2	✓

Here are some important properties of radio waves and microwaves:

- They are reflected by metal surfaces.
- They heat materials if they can make particles in the material vibrate.
- The amount of heating depends on the power of the radiation and the time that the material is exposed to the radiation.

Radio waves are produced when an alternating current flows in an aerial. They spread out and travel through the atmosphere. Another aerial is used as a detector. The radio waves produce an alternating current in it, with a frequency that matches that of the waves.

Transmitting and receiving radio waves.

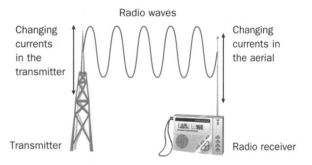

Radio waves

Changing currents in the transmitter

Changing currents in the aerial

Transmitter

Radio receiver

> Make a table of materials that reflect, absorb and transmit light, radio waves and microwaves. This will help you to remember the similarities and the differences.

Microwaves are transmitted through glass and plastics. They are absorbed by water, though how well depends on the frequency (energy) of the microwaves. Microwave ovens use a microwave frequency which is strongly absorbed by water molecules, causing them to vibrate, increasing their **kinetic energy**. This heats materials containing water, for example food. The microwaves penetrate about 1 cm into the food. Conduction and convection processes spread the heat through the food.

Microwave oven radiation will heat up our body cells and is very dangerous at high intensity because it will burn body tissue. The radiation is kept inside the oven by the reflecting metal case and metal grid in the door.

A microwave oven.

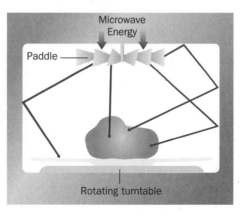

Microwave Energy

Paddle

Rotating turntable

The atmosphere

AQA	P1	✓
OCR A	P2	✓
OCR B	P1	✓
EDEXCEL	P1	✓
CCEA	P2	✓

Radio waves are transmitted through the **atmosphere** without being absorbed. Medium wavelength radio waves are reflected from the **ionosphere**, which is a layer of charged particles in the upper atmosphere.

Most microwaves are transmitted through the atmosphere, but some wavelengths are absorbed or scattered by dust and water vapour in the atmosphere and also by water droplets in rain and clouds. They pass through the ionosphere without being reflected.

How radio waves and microwaves travel.

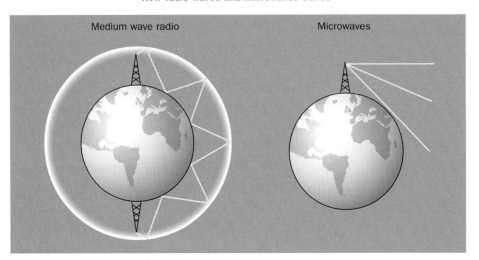

Medium wave radio   Microwaves

PROGRESS CHECK

1. Give two differences between an object you look at and the image formed on your retina.
2. What material would you use for a cooking container for use in a microwave oven? Explain your choice.
3. An 850 W microwave oven heats food for 10 seconds. How much energy in joules heats the food?
4. Why are the microwaves used in microwave ovens especially dangerous to the human body?
5. How is the eye similar to a digital camera?
6. How could microwaves that are absorbed by water droplets be used in weather forecasting?

6. If they are sent through the atmosphere between a transmitter and a receiver, the amount received will depend on whether it is raining, this could be used to track rainfall.
5. They both have a lens, small hole for light to enter, form an image on a light sensitive surface, electric signals are produced that go to the memory/brain.
4. They are absorbed by water molecules and heat them up. The human body contains a lot of water, or human cells contain water.
3. 850 W × 10 s = 8500 J
2. Glass or plastic because they do not absorb microwaves.
1. The image is smaller and upside down compared to the object.

3.3 Wireless communication

After studying this section, you should be able to:

- Explain how the atmosphere, the ionsphere and diffraction affect radio waves and microwaves.
- Describe how microwaves and radio waves are used in communicating.
- Discuss, using data and evidence, whether mobile phones are safe.
- Explain the differences between analogue and digital signals.

Radio waves

AQA	P1	✓
OCR A	P2	✓
OCR B	P1	✓
EDEXCEL	P1	✓
CCEA	P2	✓

Radio waves are suitable for **broadcasting** radio and television programmes to large numbers of people. Anyone with a receiver can tune it to the radio frequency to pick up the signal. When radio stations use similar transmission frequencies the waves sometimes **interfere** with each other. Medium wavelength radio waves are reflected from the ionosphere so they can be used for long distance communication, but not for communicating with satellites above the ionosphere.

Microwaves

AQA	P1	✓
OCR A	P2	✓
OCR B	P1, P5	✓
EDEXCEL	P1	✓
WJEC	P1	✓
CCEA	P2	✓

The transmitter and receiver for transmitting microwaves must be in line of sight (one can be seen from the other). Transmitters are positioned high up, often on tall masts. They must be close together so that hills, or the curvature of the Earth, cannot block the beam. Signals can be sent to and from **satellites**, because microwaves can pass through the **atmosphere** and through the **ionosphere**. The satellites can relay signals around the Earth which may be for television programmes, telephone conversations or monitoring the Earth, for example weather forecasting.

Satellite communication

Make sure you can compare the use of microwaves and radio waves and that you know the differences in how they behave, and how they are used. In 'compare' answers refer to both, for example microwaves will pass through the ionosphere, but medium wave radio waves do not.

Satellites in a **geosynchronous orbit** take 24 hours to orbit the Earth. The Earth rotates once in 24 hours, so these satellites stay fixed above the same point on the Earth's surface. They are then in the right position to send and receive microwave signals.

Diffraction

AQA	P1	✓
OCR A	P2, P7	✓
OCR B	P1, P5	✓
EDEXCEL	P1	✓

Diffraction is the spreading of a beam through gaps and around corners (see page 36). The maximum effect occurs when the gap has a similar size to the wavelength.

Radio waves of about 5 m are diffracted by large buildings. Radio waves of 1 km are diffracted around hills and through valleys, so they are able to reach most areas and are suitable for broadcasting. Microwave beams of a few centimetres are not spread round corners or around hills. This is why the transmitters and receivers must be in line of sight. When microwaves are transmitted from a satellite dish the wavelength must be small compared to the dish diameter to reduce diffraction. This means that, compared to radio waves, microwaves can be sent as a thin beam.

Diffraction of microwaves and radio waves

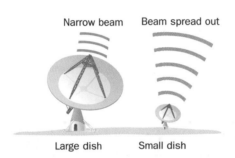

Narrow beam Beam spread out

Large dish Small dish

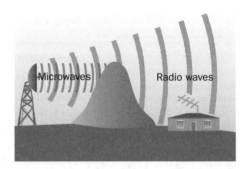

Microwaves Radio waves

Analogue and digital signals

OCR A	P2	✓
OCR B	P1	✓

A common mistake is to think that we can hear radio waves. We cannot hear any electromagnetic radiation. The radio waves are the carrier used to carry a signal that is converted into a sound wave by the receiver.

Radio waves, microwaves, infrared and visible light are all used to carry information, which can be sound, pictures or other data. The information is called the **signal**. It is added to an electromagnetic wave called the **carrier wave** so that it can be transmitted. When the wave is received the carrier wave is removed and the signal is reconstructed. There are two types of signal, **analogue** and **digital**.

> **KEY POINT**
>
> An analogue signal changes in frequency and amplitude continually in a way that matches changes in the voice or music being transmitted. A digital signal has just two values – represented as 0 and 1 (or on and off).

In digital transmissions, the signal is converted into a code of 0 and 1 values. The signal is added to the carrier wave and transmitted. After the signal is received it is decoded to recover the original signal.

Analogue and digital signals.

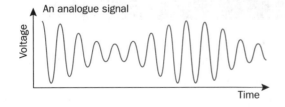

An analogue signal

Voltage

Time

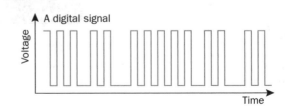

A digital signal

Voltage

Time

Both analogue and digital signals can pick up unwanted signals that distort the original signal. These unwanted signals are called **noise**. Digital signals give a better quality reception because the noise is more easily removed. They can be cleaned up in a process known as **regeneration**. The 0 or 1 values can be restored because each pulse must be a 0 or a 1. Analogue signals can be **amplified**, but the noise is amplified too.

Wireless and mobile phones

AQA	P1	✓
OCR A	P2	✓
OCR B	P1	✓
EDEXCEL	P1	✓
WJEC	P1	✓
CCEA	P2	✓

Wireless communication uses microwaves and radio waves to transmit information. The advantages of this are:

- No wires are needed to connect laptops to the internet, or for mobile phones or radio.
- Phone calls and e-mail are available 24 hours a day.
- Communication with wireless technology is portable and convenient.

> An advantage of digital signals is that they can be stored and processed by computers.

Wireless communications.

Mobile phone · Wireless laptop · Internet · Router/wireless access point · Wired desktop · Wired laptop · Wireless desktop · Wireless printer

> Remember that the danger of microwaves and other electromagnetic radiation at the low energy end of the spectrum depends on its intensity. Do not confuse this with the danger of the ionising radiations at the high energy end of the spectrum.

Mobile phones use microwave signals. The signals from the transmitting phones are reflected by metal surfaces and walls, and travel through the air to communicate with the nearest transmitter mast. There is a network of transmitter masts to relay the signals on to the nearest mast to the receiving phone.

It is unclear whether there are any long-term effects of using mobile phones. There may also be a risk to residents living close to mobile phone masts. Most people consider the risks and benefits, and decide that the benefit of using a mobile phone outweighs the risk.

Looking at data and evidence

AQA	P1	✓
OCR A	P2	✓
OCR B	P1	✓
EDEXCEL	P1	✓
WJEC	P1	✓
CCEA	P2	✓

We have only limited data about the possible dangers of mobile phones. The transmitter is held close to the user's head, so the microwaves must have a small heating effect on the brain. There is no evidence that this is dangerous. However, there has not yet been enough time to determine if there is a long-term risk. We need to collect and analyse data for an average lifetime of about 80 years before we can say whether there is any evidence of a long-term risk.

The Health Protection Agency (HPA) is made up of independent scientists, who look at the evidence of the effects of radiation on health. So far studies have not found that mobile phone users have suffered any serious ill effects. Their advice is to limit the use of handsets held close to the head, especially for young children in case there are long-term effects.

> This is an example of 'How Science Works'. It applies to all areas of physics.

Scientists do lots of studies to see if the results are **repeatable**. We can have more confidence in a study if the results been **replicated** by the scientists themselves and **reproduced** by other scientists.

Correlation between factors

AQA	P1	✓
OCR A	P2	✓
OCR B	P1	✓
EDEXCEL	P1	✓
WJEC	P1	✓
CCEA	P2	✓

A good study uses a **large sample** of thousands of people. The sample will be matched, so that scientists compare the same type of people who use mobile phones with people who don't, to see if this is a **factor** that increases the risk. For example, they might compare women, children, or men who smoke. If they find a **correlation**, they will look to see if this is due to a different **cause**. A correlation is a link between two factors.

Correlation.

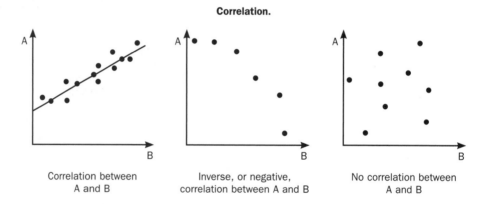

| Correlation between A and B | Inverse, or negative, correlation between A and B | No correlation between A and B |

> Remember the ice cream example so that you can use it to explain why correlation is not the same as cause.

If there is a correlation it does not always mean that one factor causes the other. For example, there is a correlation between increased sales of ice cream and an increase in hay fever, but this does not mean that eating ice cream causes hay fever. Both increase in hot sunny weather. It is important to scientists to find a **mechanism** that explains why a factor causes an **outcome**. For example, the mechanism by which radioactive materials cause cancer is the radiation ionising atoms and damaging body cells.

It is very difficult to control other factors in a study of people using mobile phones because there are so many things people do differently. Laboratory investigations should be designed to control all the factors.

Example: In an experiment to see how the distance microwaves travel in glass affects their absorption, you would make sure that these factors are kept constant:

- The intensity of microwaves from the source.
- The detector.
- The distance the microwaves travel in air before and after the glass.
- The type of glass.

PROGRESS CHECK

1. What type of electromagnetic waves are used to a) broadcast TV and b) for satellite TV?
2. Why are there often lots of microwave transmitters on the top of hills?
3. What is the difference between an analogue signal and a digital signal?
4. In communication signals, what is noise?
5. Give one advantage of wireless technology.
6. Give an example of a benefit and a risk of using a mobile phone.
7. Sketch a graph showing how ice cream sales and hay fever might be related.
8. What affect do microwaves have that might be a mechanism for causing damage to the body?
9. Why are satellite dishes larger than the wavelength of microwaves they transmit?
10. Television pictures from analogue signals often have white speckles on the picture, but digital pictures do not.
    a) What is this effect called?
    b) Why does it only happen to analogue pictures?
11. Why do scientists think that young children should use hand held mobile phones as little as possible?
12. In the example experiment on page 54, to see how distance affects absorption of microwaves:
    a) What is the factor and what is the outcome?
    b) Describe a correlation you might expect to find.
    c) What would be the effect of **not** controlling each factor?

1. a) radio waves b) microwaves
2. Need to be in 'line of sight' of the next one.
3. An analogue signal is continuously changing but digital has just two values (a labelled diagram to show this is acceptable).
4. Unwanted signals that are picked up and added to the signal.
5. We can receive phone calls and email 24 hours a day/No wires are needed to connect laptops to the internet, or for mobile phones or radio/Communication with wireless technology is portable and convenient.
6. Benefit: staying in contact with family/friends or emergency use. Risk: possible unknown long term health risk, or risk of distraction when crossing road/driving.
7. Graph showing a positive correlation – as hayfever increases sales of ice cream increase, or as sales of ice cream increase, hayfever increases.
8. A small heating effect, (or vibrate molecules).
9. So that diffraction does not make the beam spread out.
10. a) noise b) because digital signals can be cleaned up / the 0 and 1 values can be restored/ the noise/signals that produce the white specks can be removed.
11. Because they will be using them a lot over their lifetime and it is too early to be sure there are no long term effects.
12. a) factor = distance travelled in glass. Outcome = microwaves absorbed (fall in intensity detected)
    b) greater the distance the more microwaves absorbed (or lower intensity detected)
    c) They would all affect the intensity of microwaves received by the detector, so it would not be possible to tell what effect changing the distance in glass would have.

3.4 Infrared

LEARNING SUMMARY	After studying this section, you should be able to:
	• Explain properties and uses of infrared radiation.
	• Describe how optical fibres are used in communications.
	• Explain how optical fibres work.

Uses of infrared

AQA	P1	✓
OCR A	P2, P7	✓
OCR B	P1	✓
EDEXCEL	P1	✓
WJEC	P1	✓
CCEA	P2	✓

A thermogram of an elephant.

> For Edexcel you need to know that infrared radiation was discovered by Herschel.

Our skin detects **infrared** radiation and we feel **heat**. All objects emit electromagnetic radiation. The amount depends on their temperature. Hot objects glow red and very hot objects glow white hot, because they emit light as well as infrared. The hotter the object, the more electromagnetic radiation it emits and the higher the maximum frequencies. Warm objects, like radiators and human bodies, emit infrared. Night-vision goggles, cameras and thermograms use false colour so that we can 'see' the radiation.

Infrared emission, absorption and reflection depends on the surface. When infrared is absorbed the particles in the surface vibrate and gain kinetic energy. This is why infrared radiation is used for cooking. The surface of food gets hot and then the inside is heated by convection and conduction. We must be careful that intense infrared radiation does not have this effect on our skin and cause burning. Some cooking appliances, like grills and toasters, emit red light as well as infrared.

Infrared radiation is used in **remote controls** for televisions and other electronic appliances. If you look at the beam from a remote control through a digital camera (which shows up infrared signals that our eyes cannot see) you can see the flashing infrared emitting diode sending the **digital signal**. These signals cannot pass through solid objects, but sometimes reflect off walls and ceilings to operate the television.

Infrared sensors

OCR A	P2	✓
OCR B	P1	✓
EDEXCEL	P1	✓
CCEA	P2	✓

> Remember that infrared comes between microwaves and red light. Make sure you can explain the differences between them, especially their properties and uses.

Infrared sensors detect human body heat, so they are used in security systems. An infrared beam can be used as part of a burglar alarm. The burglar cannot see the beam and steps into it. This blocks the radiation reaching the sensor and triggers the alarm.

Night vision goggles and cameras work by detecting the infrared radiation emitted by objects at different temperatures. The different intensities of infrared are changed to different intensities of visible light, or often to different colours to help us see the objects even more clearly. Warm objects, like humans, other animals and cars, can be easily picked out.

Communications

AQA	P1, P3	✓
OCR A	P2	✓
OCR B	P1, P5	✓
EDEXCEL	P1, P3	✓
WJEC	P1, P3	✓
CCEA	P2	✓

> **KEY POINT**
>
> Infrared radiation and light are both transmitted along glass optical fibres by **total internal reflection**.

Total internal reflection can only happen when light is refracted at a boundary with a less dense medium, for example travelling from glass to air.

The **critical angle** is the angle of incidence for which the angle of refraction is 90°. For the glass-air boundary this is about 42°.

- When the angle of incidence is less than the critical angle the light is refracted.
- When the angle of incidence is greater than the critical angle total internal reflection occurs.

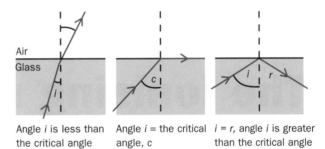

Air
Glass

Angle i is less than the critical angle   Angle i = the critical angle, c i = r, angle i is greater than the critical angle

Light travels along optical fibres.

The fibres are made with a core that has a different refractive index to the outer cladding. Radiation is reflected at the boundary as shown in the diagram.

Digital signals can be sent for long distances using **optical fibres** because glass is transparent to visible light and infrared. When the signal needs boosting, digital signals are easily regenerated. A stream of data can be transmitted very quickly. Both infrared and red light **lasers** are used as radiation sources for fibre optic communications. Lasers produce a very narrow beam of intense radiation of one colour.

Fibre optic communications

AQA	P1, P3	✓
OCR A	P2	✓
OCR B	P1	✓
EDEXCEL	P1, P3	✓
WJEC	P1, P3	✓
CCEA	P2	✓

Advantages of fibre optic communication include:

- Infrared signals in fibre optic cables experience **less interference** (they pick up less noise) than microwaves passing through the atmosphere.
- It is possible to use **multiplexing**, which is when signals are divided into sections and sent alternately so that many different signals can be sent along one fibre at the same time.

> Draw a diagram of a signal from a mobile phone travelling through a fibre optic link and a microwave link that includes a satellite. Label the diagram and it will help you to revise all the parts of microwave communication.

Lasers produce a beam with low **divergence** (does not spread out) in which all the waves have the same **frequency** and are **in phase** (in step) with each other. Lasers are also used in a compact disc (CD) player. CDs store information digitally as a series of pits and bumps (0s and 1s) on the shiny surface. A laser beam is reflected differently from the pits and bumps and a detector is used to 'read' the different reflections, reproducing the 0s and 1s of the signal.

1. How do we detect infrared radiation?
2. How does a toaster cook a slice of bread?
3. Explain why the elephant's ear in the picture on page 56 is a different colour to the rest of the body.
4. Explain how the signal from a TV remote control would be different for selecting two different channels.
5. How is light from a laser different to light from a light bulb?
6. Give an advantage of using infrared radiation and fibre-optic cable over using a microwave link.

6. Less interference. Signals can be multiplexed, many sent at once.
5. It is a thin beam (not divergent). It is one colour red – the light bulb is lots of frequencies. The waves are all in phase – in step – they are not from the light bulb.
4. It is a digital code of 0s and 1s so the pattern of 0s and 1s would be different.
3. It is cooler so the infrared radiation has a lower frequency.
2. The infrared radiation from the wire is absorbed by the molecules on the surface of the bread which gets hot. Convection and conduction then carry this heat through the rest of the bread.
1. Our skin feels it as heat.

3.5 The ionising radiations

LEARNING SUMMARY

After studying this section, you should be able to:

- State which types of radiation are ionising.
- Explain why ionising radiations are dangerous.
- Describe radioactive emissions.
- Discuss benefits and risks of using ionising radiations.

Ionising radiation

AQA	P3	✓
OCR A	P2	✓
OCR B	P2, P4	✓
EDEXCEL	P1	✓
WJEC	P1	✓
CCEA	P1, P2	✓

Gamma rays, X-rays and high energy **ultraviolet radiation** are high energy radiations which can **ionise** atoms they hit. Atoms are ionised when electrons are removed and this makes them more likely to take part in chemical reactions. If the atom is inside a living cell this can be harmful.

KEY POINT

Ionising radiation can damage or kill living cells. If the DNA in a cell is damaged it may mutate. This can cause cells to grow out of control which means that they have become cancer cells.

Whether radiation is dangerous sometimes depends on whether a person is contaminated or irradiated. Make sure you can explain the difference.

KEY POINT

There are two types of danger from all these radioactive materials:

- **Irradiation** is being exposed to radiation from a source outside the body.
- **Contamination** is swallowing, breathing in, or getting radioactive material on your skin.

A short period of irradiation is not as dangerous as being contaminated because, once contaminated, a person is continually being irradiated.

Some materials are **radioactive**. This means they randomly emit ionising radiation from the nucleus of unstable atoms. There are three types:

- Gamma rays which are electromagnetic waves.
- **Alpha particles** which are helium nuclei. They are positively charged.
- **Beta particles** which are high energy electrons from the nucleus. They are negatively charged.

Uses of ionising radiation:

- Gamma rays are used for sterilizing medical equipment and killing cancer cells.
- Alpha emitters are used in smoke detectors
- Beta emitters are used as tracers.

X-rays

AQA	P3	✓
OCR A	P2	✓
OCR B	P4	✓
EDEXCEL	P1, P3	✓
WJEC	P1	✓
CCEA	P2	✓

X-rays have high energy and pass through the body tissues. They are stopped by denser materials such as the bones and pieces of metal. They are used to image the body.

The figure shows X-rays directed towards the patient with a photographic plate placed behind. The plate darkens where the X-rays strike it and there are white shadows where the bones absorbed the X-rays.

Making an X-ray image.

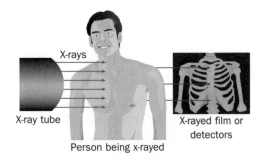

X-rays
X-ray tube
Person being x-rayed
X-rayed film or detectors

X-rays are also used for security scans of passengers' luggage. Metal items and batteries block the X-rays and show up as shadows on the screen.

Benefits and risks

AQA	P3	✓
OCR A	P2	✓
OCR B	P1, P2	✓
EDEXCEL	P3	✓
WJEC	P1	✓
CCEA	P2	✓

There are benefits to using ionising radiation, but these must be weighed against the risks. To protect people dense materials, like lead screens and concrete, are used to absorb the radiation and act as barriers. Radiation workers, such as radiographers and workers at nuclear power stations, are monitored to make sure they are not exposed to high levels of radiation. They wear protective clothes.

A patient should not receive too many X-rays. They are used when the benefit (for example, finding a broken bone) is greater than the risk. In the 1950s all pregnant women were X-rayed to check on the development and position of the baby. This caused a few cancers among children. The benefits were not greater than the risks and routine X-rays were stopped.

Radiation workers do not benefit from a dose of radiation and would be at risk from high exposure every day. This is why radiographers leave the room when a patient is X-rayed.

People tend to overestimate the risks from unfamiliar things compared to familiar things (for example they may think flying is more risky than cycling). Because ionising radiation can't be seen, people tend to think it is more dangerous than it is.

Ultraviolet radiation

OCR A	P2	✓
OCR B	P1	✓
EDEXCEL	P1	✓
WJEC	P1	✓
CCEA	P2	✓

The lower energy ultraviolet radiation that reaches the Earth's surface can cause:

- premature skin aging
- suntans
- sunburn
- skin cancer
- damage to the eyes.

> For Edexcel you need to know that Ritter discovered ultraviolet radiation.

The number of cases of skin cancer has increased in recent years as people spend more time in the Sun. Dark skins absorb more ultraviolet radiation than light skins, which are more easily damaged. To reduce the risk people should stay out of the Sun during the hottest part of the day and cover up with a hat and clothes. They should use a high protection factor sunscreen. Ultraviolet blocking sunglasses are recommended to protect the lens of the eye from damage.

> In a question about advantages and disadvantages, or about benefits and risks, you must give both to get full marks. A list of benefits cannot be awarded a mark for stating a risk.

There are benefits of spending time in the Sun, for example our skin makes vitamin D which reduces the risk of other cancers, so benefits must be weighed against risks.

Ultraviolet radiation is used to detect forged bank notes because genuine bank notes have some features that **fluoresce** in ultraviolet radiation.

PROGRESS CHECK

1. What does it mean if radiation is ionising?
2. What is the difference between being contaminated or irradiated by radioactive material?
3. Why are babies no longer routinely X-rayed in the womb?
4. Which type of electromagnetic radiation tans the skin?
5. How does leaving the room when an X-ray is taken protect a radiographer?
6. Give a benefit and a risk of sunbathing.

6. Ultraviolet causes skin to make vitamin D, but also causes skin cancer.
5. Walls absorb X-rays, so they will not be exposed.
4. Ultraviolet
3. Because the ionising radiation is damaging and the benefit does not outweigh the risk.
Irradiation is when the rays from radioactive material reach your body.
2. Contaminated is when you swallow/breathe-in/get covered by radioactive material.
1. It removes electrons from atoms making them more likely to take part in chemical reactions.

3.6 The atmosphere

LEARNING SUMMARY

After studying this section, you should be able to:

- Explain the atmospheric greenhouse effect.
- Explain the factors affecting carbon dioxide in the atmosphere.
- Discuss global warning and climatic change.
- Explain the role of the ozone layer and ultraviolet radiation.
- Explain how the ozone hole formed, and action to reduce it.

Absorption of radiation

AQA	P1	✓
OCR A	P2	✓
OCR B	P2	✓
EDEXCEL	P1	✓
WJEC	P1	✓

The Earth is surrounded by an atmosphere made up of different gases. It allows some frequencies of electromagnetic radiation from the Sun to pass through (see page 48).

> **KEY POINT**
>
> The Earth emits infrared radiation at lower frequencies. These frequencies are absorbed by some gases in the atmosphere such as **carbon dioxide**, **water vapour** and **methane**. This keeps the Earth warm and it is called the atmospheric **greenhouse effect**. Without it the Earth would be much colder – too cold for some species to survive.

The atmospheric greenhouse effect

Less global warming

More global warming

More infrared absorbed

Infrared from Earth Radiation from Sun

Less infrared from Earth Radiation from Sun
Atmosphere containing more carbon dioxide

If the power radiated away = the power absorbed the temperature will stay the same.

Earth

Earth

The greenhouse effect and global warming

OCR A	P2	✓
OCR B	P2	✓
WJEC	P1	✓

High energy infrared and visible radiation from the Sun passes through the glass into a greenhouse. Low energy infrared radiation from inside the greenhouse cannot escape through the glass. This keeps the greenhouse warm and is called the greenhouse effect.

The carbon cycle.

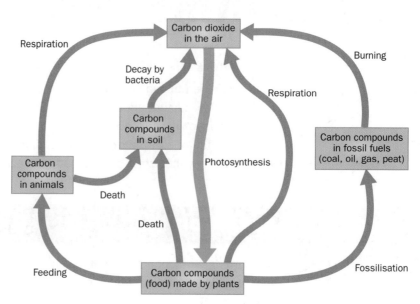

Scientists agree that **global warming** is happening. The Earth is getting warmer.

In the last 200 years the amount of carbon dioxide in the atmosphere has steadily increased. Reasons include:

- Burning **fossil fuels**.
- Clearing forests so that fewer trees are using carbon dioxide for **photosynthesis**.

For OCR B you need to know that volcanic ash in the atmosphere can cause global cooling.

Computer climate models show that human activities are increasing greenhouse gases and causing global warming, but some scientists do not agree. They do not agree about how much change is likely and what effect it will have. It could result in:

- Extreme weather conditions in some regions, such as droughts, heavy storms, flooding, very high or very low temperatures, more hurricanes, higher or less snowfall.
- Rising sea levels due to melting ice and expansion of water in oceans which may flood low lying land.
- Some regions no longer able to grow food crops, such as areas bordering deserts.

Ultraviolet in the atmosphere

| OCR A | P2 | ✓ |
| OCR B | P1 | ✓ |

Some ultraviolet radiation from the Sun reaches the surface of the Earth but, fortunately for living things, the highest energy ultraviolet radiation is stopped by the **ozone layer**. This is a layer of ozone gas in the upper atmosphere that absorbs ultraviolet radiation and chemical changes occur. Without this layer, higher frequency ultraviolet would cause more sunburn, skin cancer and cataracts.

In 1985 an ozone hole was discovered above the Antarctic. Scientists had been looking for a reduction in the ozone layer, but did not expect to find a large hole that formed so quickly. The measurements were repeated with new equipment. They were replicated by the scientists who discovered the hole and reproduced by other scientists. Where the ozone layer has been depleted, living organisms, especially animals, suffer more harmful effects from ultraviolet radiation.

The hole in the ozone layer.

A common mistake is to muddle up the effect of the ozone layer on ultraviolet and the effect of infrared on the greenhouse effect. Make sure you understand the difference.

The ozone hole

| OCR A | P2 | ✓ |
| OCR B | P1 | ✓ |

The explanation that scientists had for the ozone hole was that using gases called **CFC gases** in aerosol cans (as the propellant to force the contents of the can out of the nozzle) and as the refrigerant in refrigerators and freezers, caused the concentration of CFC gas in the atmosphere to increase. This pollution reacted with the ozone in the springtime and reduced the amount in the ozone layer. Scientists predicted that the ozone hole would grow. This prediction was correct.

International agreements, like the Montreal Protocol in 1987, have stopped the use of CFCs and other ozone depleting gases. This is a good example of the world's governments working together. The ban is having an effect and the hole is getting smaller, but the CFCs are very stable so it will be at least 50 years before the hole disappears.

PROGRESS CHECK

1. Which gases in the atmosphere contribute to global warming?
2. Why is there more carbon dioxide in the atmosphere today than there was 100 years ago?
3. Which gas absorbs ultraviolet radiation and protects living things?
4. What effect did scientists discover in the 1980s that could increase cases of skin cancer in parts of the world?
5. Most scientists agree that the climate is changing. What do some disagree about?
6. Which gases cause the hole in the ozone layer and what has been done about this problem?

6. CFCs. They have been banned by governments all over the world.
5. Whether human activities are responsible.
4. The hole in the ozone layer.
3. Ozone.
2. Because of burning fossil fuels and clearing forests.
1. Carbon dioxide, water vapour and methane.

Sample GCSE questions

1 This diagram shows how the signal is sent from one mobile phone to another using the nearest mobile phone mast to each phone to receive and resend the signal.

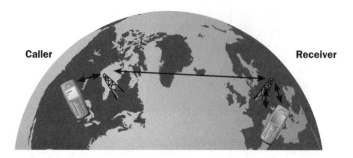

(a) What type of electromagnetic radiation do mobile phones use to send and receive signals? **[1]**

Microwaves

A mobile phone uses waves with a frequency of 900 MHz. ← MHz = Megahertz = 1 million or 10^6 Hz

(b) The wavelength of these waves in metres is calculated from

A $(3 \times 10^8) \times (900 \times 10^6)$ C $(300 \times 10^3) \times 900$
B $(3 \times 10^8) \div (900 \times 10^6)$ D $(300 \times 10^3) \div 900$

$\boxed{B}$ **[1]**

The mobile phone masts are placed on the top of hills and tall buildings, so that they are in line of sight of each other.

(c) Explain what 'in line of sight' means and why the masts have to be positioned in this way. **[2]**

Line of sight means that you can draw a straight line from one to the next without it being blocked by anything, so you would be able to see from one to the next. This is needed so that the microwaves will reach the receiving mast because they travel as a thin beam and do not spread out like radio waves.

Mobile phones are held close to the head and there have been concerns that this could be dangerous.

(d) The effect of this radiation on the brain is:

A a small heating effect   C a small ionising effect
B a small illuminating effect   D no effect

$\boxed{A}$ **[1]**

Scientific studies have been carried out, and more are being conducted, to find out whether mobile phone users are more at risk of developing cancer.

(e) Describe how scientists set up and carry out a well designed study to see if mobile phone users are more at risk of developing a brain

Sample GCSE questions

tumour. *The quality of your written communication will be assessed in this answer.* **[6]**

Scientists will set up two large (1 point) matched (1 point) samples. One will be people who use a mobile phone, and one will be people who do not (1 point) . Other factors like age, sex and other habits like smoking will be kept the same (1 point) (matched) in both groups. The larger the samples and the closer the matching the more confidence we have in the results (1 point). They will follow the groups over at least 10 years, (1 point) monitoring their mobile phone use and seeing how many people develop brain tumours in each group (1 point). If the group using phones develops more brain tumours (1 point) they will have found that mobile phones are a risk factor in developing a brain tumour.(1 point)

Scientists have not found evidence of a health risk but they still advise that young children should limit their use of mobile phones held close to their head.

(f) Explain why they give this advice. **[2]**

Mobile phones have not yet been in use for a complete lifetime of 70 or 80 years, so there has not been time to collect evidence of long term risks. Children will be using phones for many years to come.

[Total = 13]

2 To switch her TV on and off, and to change channel, Kate uses a remote control. This sends out a beam of infrared radiation. Infrared radiation is a type of electromagnetic radiation.

(a) Give two properties of electromagnetic radiation. **[2]**

1. Travels in a vacuum

2. Transverse waves

This diagram shows Kate using the remote control. She discovers that she can operate the TV by pointing the remote control at the ceiling.

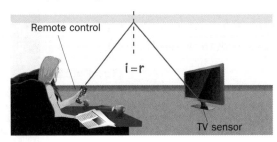

Remote control

i = r

TV sensor

(b) On the diagram, complete the path taken by the infrared radiation to operate the TV. **[2]**

Marks will be awarded depending on the number of relevant points included in the answer and the spelling, punctuation and grammar. In this question there are 9 relevant points so 8 or 9 points with good spelling punctuation and grammar will gain full marks.

For the last point an alternative is to say that children's skulls are thinner so the exposure would be higher.

There are other properties which are also correct, for example, travel at 3 x 10⁸ m/s in a vacuum, are oscillations of an electric and magnetic field.

Straight line from remote to ceiling, straight line from ceiling to TV, angle of incidence = angle of reflection. Use a ruler. There will be a tolerance allowed for the angle, but write that i = r in case your drawing is not accurate enough.

Sample GCSE questions

To operate the TV the remote control transmits a digital signal.

(c) What is the difference between an analogue signal and a digital signal? **[2]**

A digital signal has two values, often represented as 0 and 1. An analogue signal is continuously varying.

You can answer by referring to a labelled diagram as on page 52.

Infrared radiation is also used to transmit phone signals, but not through the atmosphere.

(d) What are used, together with infrared radiation, to transmit phone signals? **[1]**

optical fibres

The atmosphere absorbs infrared radiation and this helps to keep the Earth warm.

(e) What is this effect called? **[1]**

The atmospheric greenhouse effect.

(f) There has been a large change in the amount of a gas that absorbs infrared radiation in the atmosphere over the last one hundred years.

(i) What is this gas called? **[1]**

Carbon dioxide

(ii) Explain why the amount of this gas has changed and what effect this may have on the Earth. *The quality of your written communication will be assessed in this answer.* **[6]**

Plants take carbon dioxide from the atmosphere by photosynthesis (1 point). So trees reduce the amount and fossil fuels contain a lot of carbon (1 point). In the last one hundred years there has been a lot of deforestation (1 point) and burning of fossil fuels (1 point), so the amount of carbon dioxide in the atmosphere has increased (1 point). This increase causes global warming (1 point) and so human activity may be responsible (1 point) for the global warming effect which is causing the climate of the Earth to change (1 point).

[Total = 15]

Marks will be awarded depending on the number of relevant points included in the answer and the spelling, punctuation and grammar. In this question there are 8 relevant points so 6 to 8 points with good spelling punctuation and grammar will gain full marks.

Exam practice questions

1 List these electromagnetic waves in order of increasing energy. **[2]**

infrared   **radio waves**   **ultraviolet**   **visible light**

....

2 Explain why microwaves are suitable for satellite TV signals but radio waves are not. **[1]**

....

3 Describe how the eye is similar to the digital camera. **[3]**

....

....

4 Describe how microwaves that are absorbed by water droplets can be used in weather forecasting. **[2]**

....

....

5 How does light travel along optical fibres? **[1]**

A by diffraction

B by dispersion

C by refraction

D by total internal reflection

☐

6 Compare cooking with a microwave oven and a grill by giving one similarity and one difference. **[2]**

....

....

7 In a village all the residents can receive long wave radio broadcasts but not mobile phone signals. A scientist says this is a diffraction effect. Explain what the scientist means, and how it accounts for the difference. **[3]**

....

....

....

Exam practice questions

8 In the 1980s scientists discovered a hole in the ozone layer. Explain how this was caused, its effects, and what has been done to reduce it. *The quality of your written communication will be assessed in this answer.* **[6]**

....

....

....

....

....

....

9 This is a list of statements about electromagnetic radiation. Write **T** for the **true** statements and **F** for the **false** statements. **[6]**

(a) Gamma rays have higher energy photons than microwaves. ....

(b) High energy ultraviolet radiation is ionizing. ....

(c) The intensity of the radiation does not depend on the energy of the photons. ....

(d) Microwaves are reflected by glass. ....

(e) Infrared radiation and light travel along glass fibres by being diffracted. ....

(f) X-rays pass through soft tissues but are absorbed by bone. ....

4 Beyond the Earth

The following topics are covered in this chapter:

- **The Solar System**
- **Space exploration**
- **A sense of scale**
- **Stars**
- **Galaxies and red shift**
- **Expanding Universe and the Big Bang**

4.1 The Solar System

LEARNING SUMMARY

After studying this section, you should be able to:

- Describe the objects in the Solar System.
- Explain how gravity keeps planets in elliptical orbits.
- Compare the geocentric and heliocentric models of the Solar System.
- Describe NEOs and evidence for them colliding with Earth.

The Solar System

OCR A	P1	✓
OCR B	P2	✓
EDEXCEL	P1	✓
WJEC	P1	✓
CCEA	P2	✓

The **Solar System** was formed over a very long time from clouds of gases and dust in space, about **five thousand million years ago**.

The **planets** orbit the **Sun**, which is the star at the centre of the Solar System. There are eight planets and Pluto, which is a dwarf planet. Some planets are orbited by one or more **moons**.

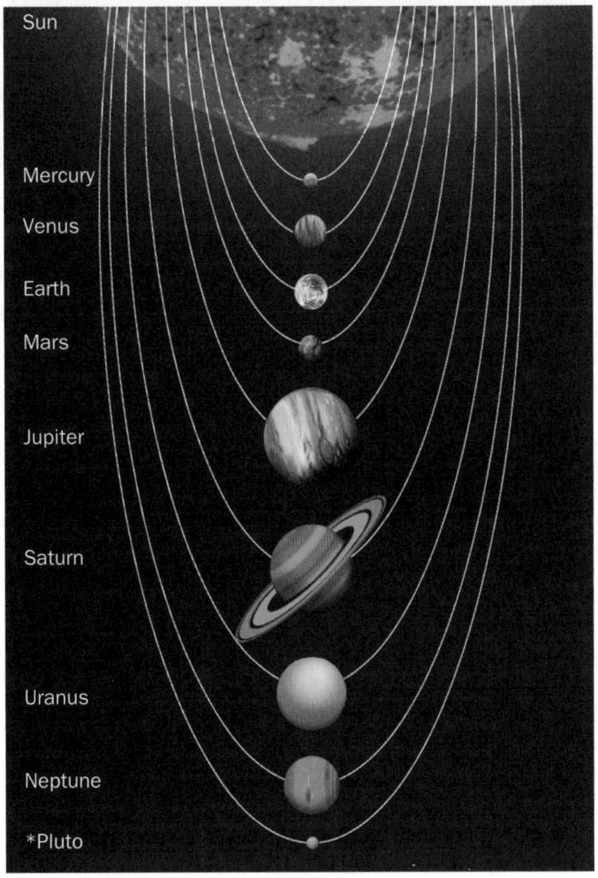

> To remember the order of the planets and Pluto (Mercury, Venus, Earth, Mars, Jupiter, Saturn, Uranus, Neptune and Pluto) use a mnemonic like:
>
> <u>My</u> <u>V</u>ery <u>E</u>asy <u>M</u>ethod <u>J</u>ust <u>S</u>peeds <u>U</u>p <u>N</u>aming <u>P</u>lanets

An object which orbits another is called a **satellite**. Our **Moon** is a natural satellite of the **Earth**.

Asteroids are rocks, up to about 1 km in diameter, that orbit the Sun. These

Sun
Mercury
Venus
Earth
Mars
Jupiter
Saturn
Uranus
Neptune
*Pluto

have been around since the formation of the Solar System. Most of these are between Mars and Jupiter.

Jupiter is the largest planet and there is a large gravitational force towards it. This has prevented the formation of a planet between Mars and Jupiter, in the asteroid belt.

There are many **comets**. Some take less than a hundred years to orbit the Sun, while others take millions of years. They are made of ice and dust. Most have a nucleus of less than about 10 km, but this vapourises and becomes a cloud thousands of miles across when the comet is close to the Sun. Comets spend most of their time far from the Sun – much further away than the dwarf planet Pluto.

> Do not confuse asteroids, meteors, comets and meteorites.

Meteors, or shooting stars, are caused by dust and small rocks, usually from a comet. When the Earth passes through this debris the rocks fall through the atmosphere. They heat up and glow. Any pieces that land on the Earth are called **meteorites**.

Near Earth Objects (NEOs) are **comets** and **asteroids** in an orbit that brings them close to the Earth. Some of them could one day collide with Earth. The **craters** on the Moon are evidence of collisions in the past. Craters on the Earth have been mostly eroded away, but layers of unusual elements in rocks and sudden changes in fossil numbers between adjacent layers of rock are evidence of collisions in the past.

> NEOs are difficult to observe because they are small and dark.

Surveys by **telescopes** try to observe and record the paths (trajectories) of all NEOs. They can be monitored from Earth or by satellite. We can make sure that we have advance warning of a collision. The idea of deflecting a NEO using explosions is being considered. At present scientists are collecting information about the composition and structure of NEOs. The possibility of destroying one is still only in the planning stage.

Gravity and orbits

AQA	P3	✓
OCR B	P2, P5	✓
CCEA	P1, P2	✓

The **force of gravity** is an attractive force between the Sun and a planet. It keeps the planet moving in **orbit** around the Sun. For an object to move in a circle there must be a force on it towards the centre of the circle. This force is called the **centripetal force**. If an object moves closer to the Sun the gravitational force on it will increase and it will speed up.

Comets have very **elliptical orbits**. They are kept in orbit by the gravitational force of attraction to the Sun, but their distance from the Sun changes as shown in the diagram. The force on the comet is largest close to the Sun, where the distance is smallest. The speed of the comet is much greater close to the Sun.

A comet's orbit.

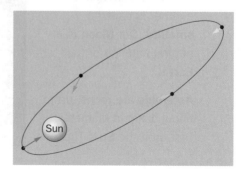

Models of the Solar System

| EDEXCEL | P1 | ✓ |
| CCEA | P2 | ✓ |

In Ancient Greece scientists thought that the Earth was stationary and that the Sun, Moon and planets moved around the Earth. This is known as the **geocentric** model. As scientists made more accurate and detailed observations of the Solar System they noticed that the planets moved in complicated patterns. A simpler explanation was that all the planets, including Earth, orbited the Sun. This is called the **heliocentric** model.

The geocentric model of the Earth at the centre of the Universe, developed by Ptolemy, is also called the **Ptolemaic model**. **Copernicus** was the first person to write down a suggested heliocentric model of the Solar System. Galileo used a new telescope to observe moons orbiting Jupiter which added to the evidence that everything did not have to orbit the Earth. This new evidence was not accepted for many years because it did not agree with religious beliefs at the time.

PROGRESS CHECK

1. Which is the largest planet?
2. What is the difference between an asteroid and a comet?
3. Describe the heliocentric model of the Solar System.
4. What is the nearest star to Earth?
5. What force keeps a comet moving around the Sun?
6. What are the main differences between the orbit of a comet and the orbit of the Earth?

1. Jupiter.
2. An asteroid is a rock that orbits the Sun between Mars and Jupiter. A comet is a rock and ice that orbits the Sun in a very elliptical orbit, spending most of the time further away from the Sun than Pluto.
3. The Sun is at the centre of the Solar System. The planets orbit the Sun and moons orbit the planets.
4. The Sun.
5. Gravity.
6. The orbit of a comet is very elliptical going close to the Sun and further away than Pluto. The orbit of the Earth is almost circular and it stays roughly the same distance from the Sun.

4.2 Space exploration

LEARNING SUMMARY

After studying this section, you should be able to:

- Describe and compare ways of finding out about the Solar System.
- Explain advantages and disadvantages of manned and unmanned space missions.
- Explain how scientists think the Moon was formed.

Exploring the Solar System

OCR A	P7	✓
OCR B	P2	✓
EDEXCEL	P1	✓
CCEA	P2	✓

Methods of finding out about the Solar System and what is beyond it are:

- To use telescopes.
- To send unmanned space probes.
- To send manned spacecraft.

Beyond the Solar System, using a telescope is the only way to study the Universe, because of the huge distances and time needed to travel them. Telescopes are also useful for study of the Solar System. They can be positioned on Earth where they are easier to maintain and repair, but where they must look through the **atmosphere**. Some telescopes are positioned high up on mountain tops to get above the clouds, dust and pollution, and away from city lights. Telescopes in orbit avoid these problems and have a much clearer view of the stars. Launching a telescope into orbit is more expensive than building one on Earth. There are telescopes designed to use all types of **electromagnetic waves**, including **radio telescopes**, which observe radio waves from space. As X-rays do not pass through the atmosphere, **X-ray telescopes** are always placed in orbit.

> Remember if you are asked to compare telescopes on Earth and in orbit, or manned and unmanned space missions, that you must compare the two. Saying that telescopes in orbit are expensive will gain no marks. You must say that telescopes in orbit are more expensive than Earth based telescopes.

Hubble space telescope

A radio telescope

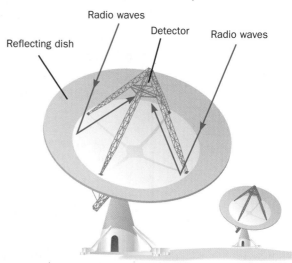

Radio waves

Detector

Radio waves

Reflecting dish

Unmanned missions

OCR B	P2	✓
EDEXCEL	P1	✓
CCEA	P2	✓

Many unmanned space probes have been launched. NASA spacecraft and Mars rovers have studied Mars. We have lots of information about Mars without humans travelling there.

Unmanned spacecraft can operate unharmed in a lot of conditions that would kill humans. By using remote sensors and computers they can send back information on:

- temperature
- magnetic field
- radiation
- gravity
- gases in the atmosphere
- composition of the rocks
- appearance of the surroundings (using TV cameras).

Manned missions

OCR B	P2	✓
EDEXCEL	P1	✓
CCEA	P2	✓

Spacecraft have put men and women in orbit and on the Moon. There are plans to send humans to Mars, but the difficulties of space missions to other planets are extreme. These include:

- Enough food, water oxygen and fuel must be carried for the entire trip, including delays.
- The distance means that emergency supplies could not be sent in time to be of use.
- An artificial atmosphere must be set up inside the spacecraft, and levels of carbon dioxide and oxygen monitored.
- Interplanetary space is very cold. There must be heating to keep astronauts warm. In direct radiation from the Sun the spacecraft will heat up, so cooling is required.
- There will be very low gravity during the journey. Bones lose density and muscles waste away under these conditions. A special exercise programme is needed using exercise machines.
- Astronauts must be shielded from cosmic rays and from the radiation released from the Sun when there is a solar flare.

Comparing manned and unmanned missions

OCR B	P2	✓
CCEA	P2	✓

Is it best to send unmanned space probes or a manned mission to find out about Mars?

Points to consider are:

- The costs of unmanned missions are less.
- There are no lives at risk if something goes wrong with an unmanned mission.
- Everything must be planned before an unmanned mission leaves – no adjustments, repairs or changes can be made unless they are programmed into the computers and can be done remotely.
- Most people are more interested in manned missions. Unmanned missions do not inspire people in the way that manned missions do.

The Moon

OCR B	P2	✓

The current theory of how the Moon was formed is that the Earth collided with another planet, about the size of Mars. Most of the heavier material of the other planet fell to Earth after the collision and the iron core of the Earth and the other planet merged. Some less dense material was thrown into orbit and formed the Moon.

Evidence for this theory is:

- Samples of Moon rocks brought back to Earth by astronauts show that Moon rocks are the same as Earth rocks – unlike rocks from other planets and moons.

- The Moon is made of less dense rocks – it only has a small iron core, unlike other planets, moons and asteroids.
- The Moon's rocks are igneous although there has been no recent volcanic activity.

4.3 A sense of scale

LEARNING SUMMARY

After studying this section, you should be able to:

- State approximate sizes and distances of objects in the Universe.
- Use the light year as a unit of distance.
- Explain how brightness and parallax can be used to measure the distance to stars.

The light year

OCR A	P1	✓
OCR B	P2	✓
EDEXCEL	P1	✓
WJEC	P1	✓
CCEA	P2	✓

To remember the increasing size; A, (no B), C, D, E. (asteroid, comet, dwarf Pluto, Earth).

A **light year** is the distance light travels in a year. Light travels through space, which is a vacuum, at 300 000 km/s. Light from the Sun arrives on Earth after about 8 minutes. After a year, light from the Sun will have travelled a distance = 300 000 km/s × (the number of seconds in a year). This distance is about 9.5×10^{12} m – about nine and a half million million kilometres. (You don't need to remember this number.)

Distances to stars and galaxies are so large that we measure them in light years. The nearest star to Earth is the Sun. The second nearest is about 4 light years from Earth.

Increasing size	Object	Size or distance	Approximate age
	Asteroids and meteors	Usually less than 1 km diameter	
	Comets	Usually less than 15 km diameter	
	Dwarf planet, Pluto		
	The Moon	Smaller than Earth (diameter is about a quarter of the Earth's)	4500 million years
	Earth	Diameter 12 760 km	4500 million years
	Largest planet, Jupiter	Diameter over 10 times the Earth's	
	Sun	Diameter over 100 times the Earth's	5000 million years
	Earth's orbit		
	Solar System	Diameter about ten thousand million km (10 000 000 000 km) Diameter about 10 light years	4600 million years
	Next nearest star (after the Sun)	About 4 light years	
	Milky Way galaxy	Diameter about 100 000 light years	
	Milky Way to nearby galaxies	About 100 000 light years	
	The observable Universe	About 14 thousand million light years	About 14 thousand million years

Looking back in time

OCR A   P1 ✓
CCEA   P2 ✓

> The light year is a unit of distance, not a unit of time.

When you look through a telescope and see a star 100 light years away, what you see – the light entering your telescope – left the star 100 years ago. So looking at very distant planets is like looking back in time. Distant objects look younger than they really are. If an alien 950 light years away could look at Earth through its telescope on the right day in 2016 it could watch the Battle of Hastings in 1066.

Brightness

OCR A   P1, P7 ✓

A very bright star may look bright because:

- It is bigger than other stars.
- It is hotter than other stars.
- It is closer than other stars.

> **KEY POINT**
>
> The brightness can be used to work out the distance to a star. The further away the star, the dimmer it is.

If there are reasons to believe stars are similar (for example, they are the same colour) then a difference in **brightness** can be used to measure the distance.

Parallax

You can see the **parallax** effect from a train or car window; objects close to you seem to move and change position more quickly, when compared to distant objects.

When two observations of the night sky are made at different times of the year, the Earth has moved. A star that is close to Earth will have changed its position when compared to distant stars in the background. The amount the nearby star has moved is used to calculate how far it is from Earth.

This method uses the fact that the Earth orbits the Sun. One observation of the night sky is made and then a second six months later. The Earth has changed its position in space by a distance equal to the **diameter** of its orbit. However, most stars are so far away that the distance moved is too small to measure.

Parallax.

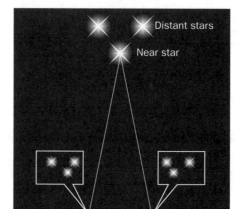

In 6 months, a close star appears to move

Another difficulty for astronomers is the amount of **light pollution** from the Earth. All the light from our cities shines into the night sky and makes it difficult to pick up the very weak signals from distant stars and galaxies.

The difficulties in making observations lead to uncertainty in our measurements of the distances to stars and galaxies.

PROGRESS CHECK

1. List the following objects in order of increasing size: a comet, the Universe, the Earth, the Moon, the Milky Way, the Sun.
2. How long does it take light to cross the Solar System?
3. A student notices that looking out of a bus window a lamp post appears to move against the background more than a bus stop sign. Which is closer?
4. Why does the Sun look different to other stars?
5. Explain one problem with using the brightness of a star to work out its distance from Earth.
6. Explain why the parallax method can only be used for nearby stars.

6. Distant stars don't appear to move when the Earth changes its position in 6 months.
5. You don't know how bright the star really is.
4. It's much closer.
3. Lamp post.
2. About 10 years.
1. Comet, Moon, Earth, Sun, Milky Way, Universe

4.4 Stars

LEARNING SUMMARY

After studying this section, you should be able to:

- Explain how we get information about stars.
- Describe how stars are formed.
- Describe main sequence stars.
- Explain what happens to stars like the Sun.
- Explain what happens to very massive stars.

How we know about stars

OCR A	P1, P7	✓
WJEC	P1	✓
CCEA	P2	✓

All the information we have about objects outside the Solar System comes from observations made with telescopes. What we know depends on the **electromagnetic radiation** from the stars and galaxies.

All around the Universe there are new stars being formed, **main sequence stars** in the stable part of their 'lifetime' and older stars coming to the end of their time as a star. By observing all of these scientists have worked out the 'life history' of stars.

To find out what stars are made of scientists look at the **spectral lines** (see page 80) in the spectrum of radiation from a star. These are used to identify the elements present in the star.

The lifetime of a star

AQA	P2	✓
OCR A	P1, P7	✓
EDEXCEL	P1	✓
WJEC	P3	✓

Formation:

- Stars begin as large clouds of dust, hydrogen and helium called an **interstellar gas cloud** or a **nebula**.
- Gravity makes the nebula contract, this makes it heat up and it becomes a **protostar**.
- When it is hot enough **hydrogen nuclei fuse** together to form **helium nuclei**. This is called **nuclear fusion**. Energy is released as light and other electromagnetic radiation. A **star** has formed.

Stable lifetime, fusing hydrogen:

- The star is one of a large number of **main sequence stars** like our Sun. Stars spend a long time fusing hydrogen. Our Sun will do this for ten thousand million years.
- What happens when a star has fused most of its hydrogen depends on the mass of the star.

Final stages of a small star like our Sun:

- The star cools, becoming redder and expands to form a **red giant**. The core contracts and **helium nuclei fuse** to form carbon, oxygen and nitrogen.
- After all the helium has fused the star contracts and the outer layers are lost. As these outer layers move away they look to us like a disc that we call a **planetary nebula**.

- The remaining core becomes a small, dense, very hot **white dwarf**. The star will then cool over a very long time and become a **brown** or **black dwarf**.

The life cycle of a star of mass similar to our Sun.

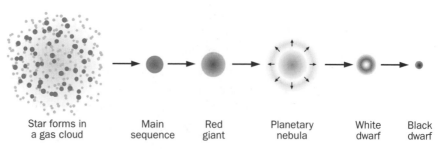

| Star forms in a gas cloud | Main sequence | Red giant | Planetary nebula | White dwarf | Black dwarf |

Final stages of a more massive star:

- The star cools and expands to form a **red supergiant**. The core contracts and **fusion** in the core forms carbon and then heavier elements up to the mass of iron.
- When the nuclear fusion reactions are finished the star cools and contracts, which heats it again until it explodes. The explosion is called a **supernova** and is the largest explosion in the Universe. All the elements heavier than iron that exist naturally on planets were created in supernova explosions.
- The core is left as a **neutron star**. It is very dense. It has a large mass, but a very small volume. If the star is very massive, the core is left as a **black hole**.

The life cycle of a more massive star.

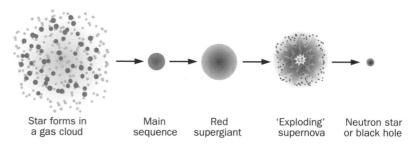

| Star forms in a gas cloud | Main sequence | Red supergiant | 'Exploding' supernova | Neutron star or black hole |

Fusion in stars

AQA	P2	✓
OCR A	P1, P7	✓
EDEXCEL	P2	✓
WJEC	P3	✓
CCEA	P2	✓

The particles of dust and gases in the interstellar gas clouds are attracted by gravitational forces between the particles. This is why the cloud contracts. As the core gets hotter the atoms collide at high speed, losing their electrons. The temperature has to be high enough to force the nuclei close together for **fusion** to occur. When light nuclei fuse they release energy.

In a main sequence star the high pressure in the core is balanced by the gravitational forces. The largest stars fuse hydrogen the most quickly so, surprisingly, more massive stars have shorter lifetimes. Only the most massive stars can cause larger nuclei to fuse, forming elements like magnesium and silicon, but even they cannot form nuclei more massive than iron.

Black holes are so dense that even light cannot escape from their strong gravitational fields.

4.5 Galaxies and red-shift

LEARNING SUMMARY

After studying this section, you should be able to:

- Explain what spectral lines are.
- Explain how spectral lines can be used to identify elements in stars.
- Explain what a red-shift is.
- Describe what Hubble discovered about distant galaxies.

Galaxies

AQA	P1	✓
OCR A	P1, P7	✓
OCR B	P2	✓
EDEXCEL	P1	✓
WJEC	P1	✓
CCEA	P2	✓

Galaxies are collections of thousands of millions of stars. There are thousands of millions of galaxies in the Universe. All the galaxies are moving away from us, and from each other.

Our Sun is a star in the **Milky Way galaxy**. To see the Milky Way from Earth (it looks like a milky strip of stars in the sky) you need to be far away from the light pollution of towns and cities. The Milky Way galaxy is shaped like a flat disc with spiral arms. Between galaxies there are empty regions that are hundreds of millions of light years across.

> Remember 'thousands of millions' of stars in a galaxy *and* galaxies in the Universe.

How science works

OCR A	P1, P7	✓
OCR B	P2	✓
EDEXCEL	P1	✓

The **scientific community** shares ideas in meetings called **conferences** and in **published journals**. They evaluate each other's work. This process is called **peer review**.

In 1920 scientists had questions about our galaxy and the Universe.

- What were **spiral nebulae?**
- Was our galaxy, the Milky Way, the only galaxy?
- How big was the Milky Way galaxy?

Two scientists held a **Great Debate**.

Harlow Shapley argued that the Milky Way was the only galaxy. He thought spiral nebulae were gas clouds. **Heber Curtis** argued that spiral nebulae were other galaxies at great distances away from the Milky Way.

There was not enough evidence to prove who was right. At the time most scientists thought that Shapley made a better case.

In 1924 Edwin Hubble used a new telescope and a new method of measuring distances. This new evidence showed that the distances to spiral nebulae were much greater than the size of the Milky Way.

He published his results and they were **evaluated** and **reproduced** by other scientists. Distances to more spiral nebulae were measured. They were outside the Milky Way. The scientific community accepted Curtis was right.

New explanations

OCR A	P1	✓
OCR B	P2	✓
EDEXCEL	P1	✓

When new data disagrees with an accepted explanation, scientists do not immediately give up their accepted explanation. They wait until they have an explanation that fits the data better before they give up the accepted explanation.

Absorption spectra and red-shift

AQA	P1	✓
OCR A	P1	✓
OCR B	P2	✓
EDEXCEL	P1	✓
WJEC	P1	✓
CCEA	P2	✓

Spectral lines are lines in the spectrum of radiation from a hot object like a star. Dark lines are wavelengths that have been **absorbed** by the atoms of gases the radiation has passed through, usually in the outer, cooler parts of the stars. Bright lines are wavelengths emitted by the atoms of hot gases.

Each **element** has a different set of absorption lines, so the lines can be used to identify the elements present in the stars. The **absorption spectrum** has been described as a 'fingerprint' of the element. This is because each element has a unique pattern of absorption lines.

Absorption spectrum of sunlight compared with hydrogen and sodium.

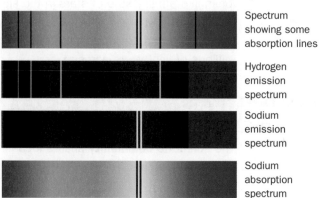

Spectrum showing some absorption lines

Hydrogen emission spectrum

Sodium emission spectrum

Sodium absorption spectrum

For visible light, red has the longest wavelength and violet the shortest. If the wavelength is longer than expected this is called a **red-shift**. If a wavelength is shorter than expected, this is called a blue-shift.

If a source of wave is stationary, the waves move out from the source at the same rate in all directions. If the source of waves is moving the wave is either stretched or squashed, depending on which way the source is moving. In the second diagram below, the source of waves is moving away and the wavelength appears longer. If the source were moving towards the observer, the wavelength would appear shorter.

> A red-shift is a shift towards the red end of the spectrum. Infrared radiation would be shifted towards microwaves not towards red light. A blue shift means the radiation source is moving towards the observer.

Waves from a moving source.

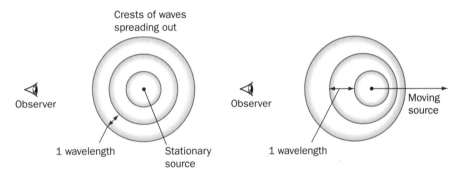

A red-shift in the light from a star shows that the distance between us and the star is increasing. The bigger the red-shift the faster the star is moving away. Hubble discovered:

KEY POINT

- The light from all the **distant galaxies** is red-shifted – so they are all moving away from us.
- The further away the galaxy, the bigger the red-shift, so the faster the galaxy is moving away.

Red-shift and the Doppler Effect

AQA	P1	✓
OCR A	P1	✓
OCR B	P2	✓
EDEXCEL	P1	✓
CCEA	P2	✓

The change in wavelength, because a source of waves is travelling towards or away from an observer, is called the **Doppler Effect**. When the wavelength increases the frequency decreases. This is why a siren drops in frequency as it travels away and increases in frequency as it approaches.

Light from stars and galaxies is red-shifted or blue-shifted depending on whether they are moving towards us or away from us. The shift moves all the dark spectral lines by the same amount.

PROGRESS CHECK

1. What do we know about the movement of distant galaxies?
2. What is 'peer review'?
3. From the absorption spectra, is sodium vapour present in the Sun?
4. If the Sun was moving away from us which way would the spectral lines move?
5. Light from the Andromeda galaxy is blue-shifted. What does this tell you about the galaxy?
6. What information can astronomers get from line spectra?

4.6 Expanding Universe and the Big Bang

LEARNING SUMMARY	**After studying this section, you should be able to:**
	• Explain how red-shift implies an expanding Universe.
	• Explain how the expanding Universe implies a Big Bang.
	• Describe how scientists develop new theories.
	• Describe the evidence for the Big Bang theory.

The expanding Universe

AQA	P1	✓
OCR A	P1	✓
OCR B	P2	✓
EDEXCEL	P1	✓
WJEC	P1	✓
CCEA	P2	✓

Hubble made two important observations:

• The light from all the distant galaxies is red-shifted.
• The further away the galaxy the bigger the red-shift.

KEY POINT

This means:

• All the distant galaxies are moving away from us.
• The further away the galaxy, the faster it is moving away.

We would not see these patterns in the red-shifts just because we, or the galaxies, are moving through space, but it is what we would see if space was expanding.

The difference between a moving star and expanding space.

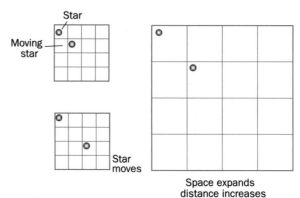

This is why scientists think we live in an **expanding Universe**. The Universe is everything that exists. There is nothing outside the Universe – not even empty space.

To help you to remember about space expanding and the two effects that Hubble observed, imagine blowing-up a balloon with galaxies drawn on it. The galaxies drawn on the balloon will all spread out, moving away from each other and those furthest apart will separate more quickly because the plastic surface is expanding.

Space is expanding.

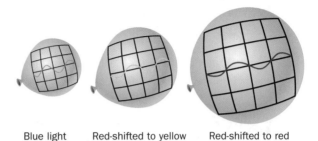

Blue light   Red-shifted to yellow   Red-shifted to red

Cosmological red-shift

AQA	P1	✓
OCR A	P1	✓
OCR B	P2	✓
EDEXCEL	P1	✓
WJEC	P1	✓
CCEA	P2	✓

All of the space in the Universe is expanding, but we do not notice this in the Solar System. This is because gravity is an attractive force that stops mass spreading out when space expands.

Stars and the nearest galaxies may show red-shifts or blue-shifts because they are moving through space and these are sometimes called Doppler shifts. Distant galaxies show red-shifts because space is expanding, so these are sometimes called cosmological red-shifts.

Spectral lines from galaxies.

Moving towards you: blue-shift

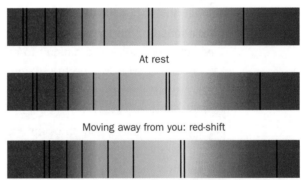

At rest

Moving away from you: red-shift

The Big Bang

AQA	P1	✓
OCR A	P1	✓
OCR B	P2	✓
EDEXCEL	P1	✓
WJEC	P1	✓
CCEA	P2	✓

KEY POINT

The fact that the Universe is expanding is strong evidence for the fact that it started as a small point. Starting from today and working backwards scientists calculate that the Universe began with a 'Big Bang' about 14 thousand million years ago. This is called the **Big Bang Theory**.

The Big Bang Theory started as a **hypothesis** – a suggested explanation thought up creatively to account for the data. Scientists then used it to make a **prediction**. They said that the Big Bang would have produced radiation that, by now, would be found in the microwave region of the spectrum. It would come from all parts of the Universe. Scientists called this the **cosmic microwave background radiation (CMBR)**.

Cosmic microwave background radiation

AQA	P1	✓
OCR B	P2	✓
EDEXCEL	P1	✓
WJEC	P1	✓
CCEA	P2	✓

Scientists began searching for the CMBR. In 1965, two scientists discovered it accidently. They were using a radio telescope and could not account for an annoying microwave signal that seemed to come equally from all directions. It was the CMBR. The Big Bang Theory was the only theory that could explain this.

Do not confuse the CMBR with radioactive background radiation, which is ionising radiation from radioactive materials.

The steady state theory

| EDEXCEL | P1 | ✓ |

The **Steady State Theory** is an alternative theory that suggests the Universe is expanding, but has always looked the same. It will continue to look the same, because new matter is being created at various places. There were already some problems with data that could not be explained by the Steady State Theory and the discovery of the CMBR led to the general acceptance of the Big Bang Theory.

Questions that can not be answered

AQA	P1	✓
OCR A	P1	✓
EDEXCEL	P1	✓
CCEA	P2	✓

Scientists may never be able to answer some questions like 'What happened before the Big Bang?' because they can't collect data from before the Big Bang.

Scientists can't answer some questions because of difficulties in collecting the data they need. For example, whether the Universe will continue to expand, or whether it will slow down and stop, or whether it will start to contract, depends on the mass of the Universe and the distance to the edge of the Universe. These are very large and difficult to measure, so scientists can not yet answer the question, 'What will be the ultimate fate of the Universe?'

PROGRESS CHECK

1. What explanation do scientists have for the red-shifts in radiation from all the distant galaxies?
2. How long ago was the Big Bang?
3. Give an example of a question science can't answer.
4. What effect was predicted by scientists if the Big Bang theory was true?
5. What led to the acceptance of the Big Bang Theory and why?

1. The Universe is expanding.
2. 14 thousand million years ago
3. What happened before the Big Bang?
4. Cosmic microwave radiation left over from the Big Bang.
5. The discovery of the cosmic microwave background radiation because other theories could not explain it.

Sample GCSE questions

1 **(a)** Write these objects in order of increasing size. The first has been done for you **[3]**

A = comet   B = diameter of Earth's orbit   C = galaxy
D = the Sun   E = the Universe

A	D	B	C	E

D anywhere before B = 1 mark, B anywhere before C = 1 mark, C anywhere before E = 1 mark.

(b) The light received from the Andromeda galaxy shows a blue-shift. What does this tell us about the Andromeda galaxy? **[1]**

It is moving towards the Earth.

(c) The diagram shows the spectra of light from two distant galaxies.

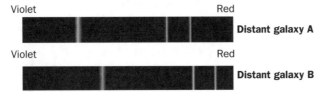

(i) Which galaxy is furthest away, A or B?

B

(ii) Explain how you can tell.

The light from galaxy B has a larger red-shift than the light from galaxy A. This means it is moving away from us faster and the further away a galaxy is the faster it is moving away. **[2]**

One mark is for identifying and explaining that galaxy B has the larger red-shift. The second mark is for explaining why this means it is further away.

(d) This graph shows the speed of distant galaxies plotted against their distance from Earth.

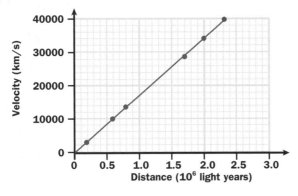

(i) Describe the relationship shown by the graph. **[2]**

The speed of a distant galaxy is proportional to its distance away from the Earth.

A straight line through the origin means that one variable is proportional to the other. This is the best way to describe the relationship. You could also say that as one variable doubles the other doubles. Identifying that the speed increases as the distance increases would gain only one of the marks.

(ii) Use the graph to estimate the speed of a galaxy that is 1.3×10^6 light years from Earth. **[1]**

22 000 km/s

Don't forget the units. It is a good idea to use a ruler to mark the value you are reading from the graph.

Sample GCSE questions

(e) Scientists have used this information to form a theory about the beginning of the Universe. Name and describe the theory, explaining a prediction made using the theory and how this lead to its acceptance by many scientists. *The quality of your written communication will be assessed in this answer.* **[6]**

The theory is the Big Bang Theory (1 point). The theory says that the Universe started to expand from a small point (1 point) about 14 thousand million years ago (1 point). Scientists predicted that there should be radiation left over from the big bang (1 point) which would be in the microwave region of the electromagnetic spectrum, called cosmic microwave background radiation (1 point). When scientists discovered the cosmic background radiation (1 point) most scientists accepted the theory.

> Marks will be awarded depending on the number of relevant points included in the answer and the spelling, punctuation and grammar. In this question there are 6 relevant points so 5 or 6 points with good spelling punctuation and grammar will gain full marks.

2 (a) How do scientists get information about which elements are present in a star? **[3]**

They look at the electromagnetic radiation from the star. They see spectral lines which are dark lines corresponding to wavelengths (or frequencies) absorbed by the elements present in the star. The element always has the same spectral lines so they can be used to identify the element.

> Remember that the electromagnetic radiation is the only information we have from stars. Many stars emit wavelengths outside the range of visible light, so em radiation is a better answer than light.

(b) Betelgeuse is a red giant star. Give two differences between Betelgeuse and our Sun. **[2]**

1. Betelgeuse is much larger than the Sun.
2. Betelgeuse is much cooler than the Sun.

> You could also say that Betelgeuse fuses helium, whereas the Sun fuses hydrogen.

(c) Draw a ring round the main sequence star that will end up as a supernova. **[1]**

The Sun   A star with smaller mass than the Sun

(A star 20 times more massive than the Sun)

(d) Explain what happens as a star leaves the main sequence and ends up as a supernovae. *The quality of your written communication will be assessed in this answer.* **[6]**

When the star has finished fusing hydrogen (1 point) to helium (1 point) it cools and expands (1 point) to form a red supergiant (1 point). The core contracts (1 point) and fusion in the core forms carbon (1 point), oxygen and nitrogen (1 point) and then heavier elements up to the mass of iron (1 point).

When the nuclear fusion reactions are finished (1 point) the star cools and contracts (1 point), which heats it again until it explodes (1 point). The explosion is called a supernova (1 point).

> Marks will be awarded depending on the number of relevant points included in the answer and the spelling, punctuation and grammar. In this question there are 12 relevant points so 10 or 12 points with good spelling punctuation and grammar will gain full marks.

Exam practice questions

1 **(a)** What is the difference between the geocentric and heliocentric models of the Solar System? **[2]**

....

....

....

(b) When Copernicus found evidence for a heliocentric model, the new evidence was not accepted for many years.

Why did it take so long for new evidence to be accepted? **[1]**

....

....

[Total = 3]

2 The current theory of how the Moon was formed suggests that a planet collided with the Earth and the less dense rocks were thrown up as the Moon. Give two pieces of evidence that support this theory. **[2]**

1. ....

2. ....

3 In the 20th century scientists collected data about the planet Mars.

(a) State whether they used

(i) Telescopes: ....

(ii) Unmanned space missions: ....

(iii) Manned space missions: .... **[1]**

(b) Describe the advantages and the disadvantages of unmanned space missions when compared to manned space missions. *The quality of your written communication will be assessed in this answer.* **[6]**

....

....

....

....

....

....

....

....

....

[Total = 7]

Exam practice questions

4 **(a)** What is a light year? [1]

....

(b) Explain why the distance to the Sun is measured in kilometers but the distance to other stars is measured in light years. [1]

....

....

[Total = 2]

5 Draw one straight line from each box on the left to match the object to its size. [3]

Object   **Size**

Object	Size
Diameter of the Earth	10 light years
Diameter of the Solar System	4 light years
Diameter of the Milky Way galaxy	12 760 km
Distance to the nearest star	100 000 light years

6 This is the spectrum of a nearby star.

Violet   Red

Star

These are the line spectra of some elements on Earth:

Violet   Red
A

Violet   Red
B

Violet   Red
C

Violet   Red
D

(a) Which of the elements are present in the star? [4]

....

Exam practice questions

When we observe light from a distant star we see a red-shift in the wavelength.

(b) What is meant by red-shift? [1]

....

(c) What does it tell us about the star? [1]

....

[Total = 6]

7 **(a)** Scientists have a theory about how the Universe was formed called the Big Bang theory. Describe the theory and explain the evidence they have for the theory. *The quality of your written communication will be assessed in this answer.* [6]

....

....

....

....

....

....

(b) At a conference some scientists present evidence that does not agree with the Big Bang theory. They say the Big Bang theory is wrong. Describe what scientists should do when they disagree about a theory. [3]

....

....

....

(c) Scientists have a theory that answers the question 'How did the Universe begin?' Explain whether they can answer the question 'Why was the Universe created?' [2]

....

....

....

[Total = 11]

5 Forces and motion

The following topics are covered in this chapter:

- Distance, speed and velocity
- Speed, velocity and acceleration
- Forces
- Acceleration and momentum
- Action and reaction
- Work and energy
- Energy and power

5.1 Distance, speed and velocity

LEARNING SUMMARY

After studying this section, you should be able to:

- Explain the difference between average speed and instantaneous speed.
- Explain that velocity and displacement have direction.
- Calculate speeds, velocities, distances and times.
- Plot and interpret distance–time graphs.

Distance, speed and velocity

AQA	P2	✓
OCR A	P4	✓
OCR B	P3	✓
EDEXCEL	P2	✓
WJEC	P2, P3	✓
CCEA	P1	✓

Speed can be measured in metres per second (m/s) or kilometres per hour (km/h).

- An athlete running with a speed of 5 m/s travels a distance of 5 metres in one second and 10 metres in two seconds.
- An athlete with a faster speed of 8 m/s travels further, 8 metres, in each second and takes less time to complete his journey.

> **KEY POINT**
>
> To calculate speed:
>
> $$\text{speed (m/s)} = \frac{\text{distance (m)}}{\text{time (s)}}$$

Example of triangle method: $\text{speed} = \dfrac{\text{distance}}{\text{time}}$

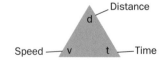

To calculate distance or time:

> Some students find using the triangle method useful to re-arrange a formula.

- Write the formula into a triangle, so that distance is 'over' time. This means putting distance at the top corner of the triangle.

- To find the **distance**, cover the word distance with your finger and look at the position of speed and time. They are side by side, so distance = speed × time.
- To find the **time**, cover the word time with your finger and distance is 'over' speed, so

$$\text{time} = \frac{\text{distance}}{\text{time}}$$

Speed and velocity

AQA	P2	✓
OCR A	P4	✓
OCR B	P3, P5	✓
EDEXCEL	P2	✓
WJEC	P2	✓
CCEA	P1	✓

Direction can be important when making calculations about a journey.

Speed does not take direction into account. Speed is calculated using the actual distance travelled no matter what direction the object is moving in. Speed is always a positive number.

If the direction travelled is taken into account, one direction can be called the positive direction and the opposite direction can be called the negative direction. The word displacement is used to measure distance with a direction. Travel in the positive direction is described as positive displacement and travel in the negative direction is described as negative displacement.

Quantities that have **magnitude** and **direction** are called **vectors**.

Velocity is speed in a given direction. Therefore, velocity is a vector. For example a dog walks in a positive direction for 5 m and then back again with a constant speed of 2 m/s, so he walks with a velocity of +2 m/s and then with a velocity of −2 m/s.

The total distance the dog has walked is 10 m, however the displacement at the end of the journey is zero.

$$\text{velocity (m/s)} = \frac{\text{displacement (m)}}{\text{time (s)}}$$

Distance-time graphs

AQA	P2	✓
OCR A	P4	✓
OCR B	P3	✓
EDEXCEL	P2	✓
WJEC	P2	✓
CCEA	P1	✓

On a **distance-time graph**:

- A horizontal line means the object is stopped.
- A straight line sloping upwards means it has a steady speed.

Distance–time graph for a cycle ride.

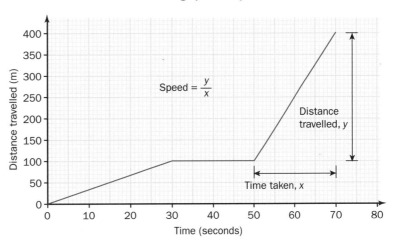

When you are doing calculations:

- **Always show your working** – if you make a mistake in the calculation you may still get some marks.
- **Do not write down** *only* **the triangle.** You will get marks for using the correct formula or equation.

The steepness, or **gradient**, of the line shows the speed:

- A steeper gradient means a higher speed.
- A curved line means the speed is changing.

Example: Distance-time graph for a cycle ride.

Between 30 s and 50 s the cyclist stopped. The graph has a steeper gradient between 50 s and 70 s than between 0 s and 30 s. The cyclist was travelling at a greater speed.

To calculate a speed from a graph, work out the gradient of the straight line section

speed $= \frac{y}{x}$ where $y = 400$ m $- 100$ m $= 300$ m and $x = 70$ s $- 50$ s $= 20$ s.

speed $= \frac{400 \text{ m}}{20 \text{ s}} = 20$ m/s.

Average speed and instantaneous speed

| OCR A | P4 | ✓ |
| CCEA | P1 | ✓ |

The **average speed** of the cyclist for the total journey shown on the graph on page 91 is:

$= \frac{\text{total distance}}{\text{total time}} = \frac{400 \text{ m}}{70 \text{ s}} = 5.71$ m/s

This is not the same as the **instantaneous speed** at any moment because the speed changes during the journey. If you calculate the average speed over a shorter time interval you get closer to the instantaneous speed.

Displacement–time graphs

OCR A	P4	✓
OCR B	P3	✓
EDEXCEL	P2	✓
CCEA	P1	✓

If the direction of travel is being considered:

- A negative displacement is in the opposite direction to a positive displacement.
- A straight line sloping upwards or downwards means steady speed.
- Upwards means a steady positive velocity, and downwards means a steady negative velocity.

Example: Displacement–time graph for a journey from home.

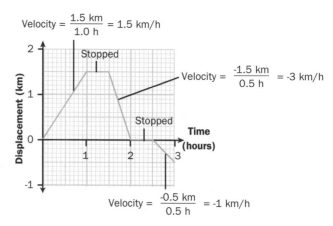

A boy starts from home (0 km) and walks to a shop, home again and then in the opposite direction to the shop.

1. What is the difference between instantaneous speed and average speed?

2. In the graph of the cycle ride, what is the speed during the first 30 seconds?

3. A car travels 288 km in three hours. Calculate the speed in km/h.

4. A car travels at a speed of 12 m/s. How long will it take to travel 1.44 km?

5. In the graph of the journey from home on page 92:
   a) What was the displacement after the first 30 minutes?
   b) When was the speed greatest?
   c) How can you tell this without doing any calculations?

1. Average speed is calculated from the distance travelled over time taken and the speed may have changed several times during the journey time. Instantaneous speed is the speed at any instant.
2. (100 m − 0 m) ÷ (30 s − 0 s) = 3.33 m/s
3. $s = 288 \text{ km} \div 3 \text{ h} = 96 \text{ km/h}$
4. $t = d \div s$ $t = 1.44 \text{ km} \div 12 \text{ m/s} = 1440 \text{ m} \div 12 \text{ m/s} = 120 \text{ s } (= 2 \text{ minutes})$
5. a) 0.75 km
   b) Between 1.5 and 2 hours (or on the way home from the shop).
   c) Because the graph is steepest at this point.

5.2 Speed, velocity and acceleration

LEARNING SUMMARY

After studying this section, you should be able to:

- Calculate acceleration.
- Plot and interpret speed–time graphs.
- Plot and interpret velocity–time graphs.
- Calculate acceleration and distance from graphs.

Acceleration

AQA	P2	✓
OCR A	P4	✓
OCR B	P3	✓
EDEXCEL	P2	✓
WJEC	P2	✓
CCEA	P1	✓

Any change of velocity is called **acceleration**. Speeding up, slowing down and changing direction are all examples of acceleration. Acceleration is the change in velocity per second. It is measured in metres per second squared (m/s^2).

Example: If a car accelerates from 0 to 27 m/s (about 60 mph) in 6 seconds the change in velocity is 27 m/s, the acceleration

$$= \frac{27 \text{ m/s}}{6 \text{ s}} = 4.5 \text{ m/s}^2$$

> **KEY POINT**
>
> $$acceleration\ (m/s^2) = \frac{change\ in\ velocity\ (m/s)}{time\ taken\ (s)}$$
>
> $$a = \frac{(v - u)}{t}$$
>
> where a is the acceleration of an object whose velocity changes from initial velocity (u) to final velocity (v) in time (t).

Example: A car accelerates from 14 m/s to 30 m/s in 8 s. The acceleration:

$$a = \frac{(30\ m/s - 14\ m/s)}{6\ s} = \frac{16\ m/s}{8\ s} = 2\ m/s^2$$

Speed–time graphs

AQA	P2	✓
OCR A	P4	✓
OCR B	P3	✓
EDEXCEL	P2	✓
WJEC	P2	✓
CCEA	P1	✓

> Always check carefully whether a graph is a speed–time graph or a distance–time graph.

Plotting the speed of an object against the time gives a graph like this:

A speed–time graph.

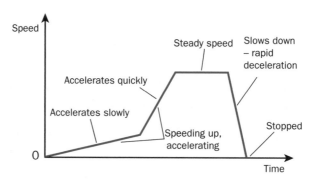

- A **positive slope** (**gradient**) means that the speed is increasing - the object is accelerating.
- A **horizontal line** means that the object is travelling at a steady speed.
- A **negative slope** (**gradient**) means the speed is decreasing – negative acceleration.
- A **curved slope** means that the acceleration is changing – the object has **non-uniform acceleration**.

On true **speed–time graphs** the speed has only positive values. On **velocity–time graphs** the velocity can be negative.

> For OCR B you need to know about speed cameras.

Tachographs are instruments that are put in lorry cabs to check that the lorry has not exceeded the speed limit and that the driver has stopped for breaks. They draw a graph of the speed against time for the lorry.

Velocity–time graphs

AQA	P2	✓
OCR A	P4	✓
OCR B	P3	✓
EDEXCEL	P2	✓
WJEC	P2	✓
CCEA	P1	✓

Graphs for a ball that rolls up a hill, slows and rolls back down, speeding up.

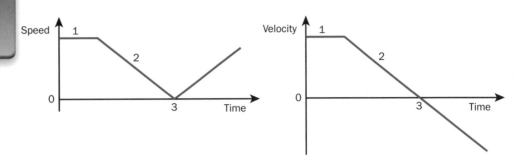

At point 3 the ball stops and starts to roll back. Its speed starts to increase again. Its velocity starts to increase, but it is now rolling downhill – so the velocity is negative.

Acceleration from a graph

The acceleration is the gradient of a velocity–time graph.

Example: A graph of velocity against time for a car journey.

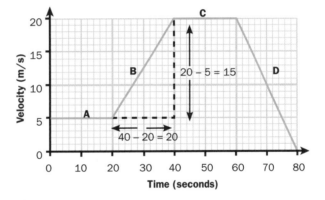

Remember the gradient is

$$\frac{rise}{run}$$

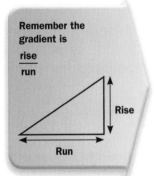

In part B of the graph the car accelerates from 5 m/s to 20 m/s in (40 − 20)s = 20 seconds. The gradient:

$$= \frac{y}{x} = \frac{15}{20} = 0.75$$

So the acceleration = 0.75 m/s^2.

Distance travelled from a graph

AQA	P2	✓
OCR B	P3	✓
EDEXCEL	P2	✓
WJEC	P2	✓
CCEA	P1	✓

On a velocity–time graph the area between the graph and the time axis represents the distance travelled.

Example: Distance travelled on a bike journey.

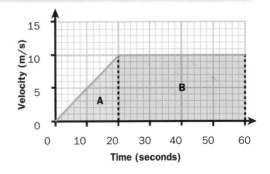

Take care with units. You may need to change minutes or hours to seconds, or kilometres to metres.

36 km/h =

$\dfrac{36 \times 1000}{60 \times 60}$ m/s

= 10 m/s

Also remember to write down the units for your answer.

The distance travelled = area under A + area under B

Area A = area of pink triangle = $\dfrac{1}{2}$ (10 m/s × 20 s) = 100 m

Area B = area of blue rectangle = 10 m/s × (60 − 20) s = 400 m

Distance travelled = 100 m + 400 m = 500 m

PROGRESS CHECK

1. A car accelerates from 0 to 24 m/s in 8 s. What is the acceleration?
2. What does a horizontal line on a speed-time graph tell you?
3. From graph of the car journey on page 95, what is the acceleration at:
   a) A
   b) D.
4. Sketch a velocity time graph for a car that speeds up in a negative direction, travels at a steady speed, and then slows down and stops.
5. From graph of the car journey on page 95, what is the distance travelled in section:
   a) C
   b) D.

5. C 20m/s × (60 s − 40 s) = 400 m
   D $\frac{1}{2}$ × 20 m/s × (80 s − 60 s) = 200 m
4. Line goes below x axis similar to last part of the graph for the rolling ball, then horizontal, then straight line sloping upwards to the x axis again.
3. A 0 m/s²
   D (0 − 20) m/s ÷ (80 − 60) s = − 1 m/s²
2. The object is travelling at a steady/constant speed.
1. 24 m/s ÷ 8s = 3 m/s²

5.3 Forces

LEARNING SUMMARY

After studying this section, you should be able to:

- Explain the difference between mass and weight.
- Calculate the resultant force on an object.
- Describe the forces on moving and stationary objects.
- Understand and use Newton's First Law of Motion.

Mass and weight

AQA	P2	✓
OCR A	P4	✓
OCR B	P3	✓
EDEXCEL	P2	✓
WJEC	P2	✓
CCEA	P1	✓

For WJEC you also need to know that mass gives an object inertia.

KEY POINT

Mass is measured in **kilograms** (kg). It is the amount of matter in an object. An object has the same mass everywhere, on the Earth, the Moon, or in outer space.

Weight is a force and is measured in **newtons** (N). Weight is the force of gravity attracting the mass towards the centre of the Earth. An object's weight changes when the force of gravity changes, for example, on the Earth, the Moon, or in outer space.

weight (N) = mass (kg) × gravitational field strength (N/kg)

Gravitational field strength close to the surface of the Earth is 10 N/kg. Objects in free fall have a constant acceleration of 10 m/s².

In outer space there is no gravity so all objects are weightless. The Moon's gravitational attraction is only one sixth of the Earth's, so objects on the Moon have only one sixth their weight on Earth.

To remember the differences between mass and weight think of a tin of beans.

- It is weightless in outer space.
- It has less weight on the Moon.
- Its mass changes if you eat the beans.

> **For OCR B you need to know that g will be slightly less on top of a mountain and more at the bottom of a mine shaft.**

Resultant and balanced forces

AQA	P2	✓
OCR A	P4	✓
OCR B	P3	✓
EDEXCEL	P2	✓
WJEC	P2	✓
CCEA	P1	✓

Forces have size and direction. On diagrams they are shown by arrows. The length of the arrow represents the size of the force. When several forces act on an object the effect is the same as one force in a certain direction. This is called the **resultant force**. Forces combine to give a resultant force. If the resultant force is zero the forces on the object are **balanced**.

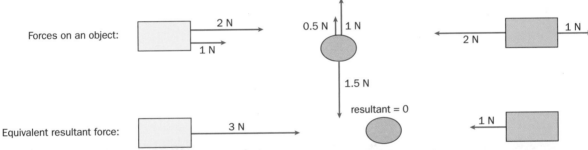

Resultant forces.

Forces on an object:

Equivalent resultant force:

> ### KEY POINT
>
> A resultant force changes the velocity of an object. This idea is known as **Newtons First Law of Motion**:
>
> *If the resultant force on an object is zero, the object will remain stationary or continue to move at a steady speed in the same direction.*
>
> When forces on an object are balanced it does not fall.

Reaction and tension are forces that can balance weight.

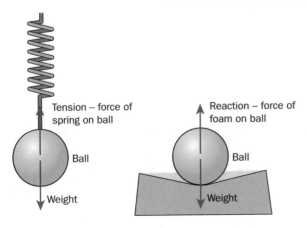

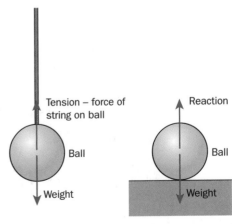

The upward force of the spring, or the foam, trying to return to their original shape, balances the weight.

Even though the changes in shape are too small for us to see, the restoring force – the tension in the string or the reaction from the floor – balances the weight.

Forces and elasticity

AQA   P2   ✓

A force acting on an object may change the shape of the object, for example in the case of the spring or the foam on page 97. Objects like these are elastic, and as the object stretches it stores **elastic potential energy**. When a spring is stretched the work done on it is equal to the energy stored. This energy is released when the spring returns to its original length.

> **KEY POINT**
>
> The extension of a spring is proportional to the force applied, provided the limit of proportionality is not exceeded.
>
> $F = k \times e$
>
> where F is the force in newtons (N), e is the extension in metres (m) and k is the spring constant in newtons per metre (N/m)

Stretching a spring, P, is the limit of proportionality.

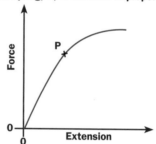

Friction

AQA   P2   ✓
OCR A   P4   ✓
OCR B   P3   ✓
EDEXCEL   P2   ✓
WJEC   P2   ✓
CCEA   P1   ✓

When one object slides, or tries to slide, over another, there is **friction**. Friction is the resistive force between the two surfaces.

Air resistance is the resistive force that acts against objects moving through the air.

Drag is the resistive force on objects moving through liquids or gases. Drag is larger in liquids.

> Drag can be reduced by shaping the object so that it is more streamlined.

These resistive forces:

- Always act against the direction of motion.
- Are zero when there is no movement.
- Increase as the speed of the object increases.

The forces on a cyclist when speeding up, at a constant speed and slowing down.

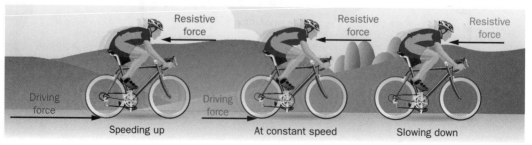

> **Friction can be reduced by lubricants.**

When the driving force is larger than the resistive force the cyclist speeds up, when they are equal he travels at a steady speed and when the resistive force is greater he slows down.

Forces when falling

AQA	P2	✓
OCR A	P4	✓
OCR B	P3	✓
EDEXCEL	P2	✓
WJEC	P2	✓

The resultant force on a skydiver changes.

Falling objects reach a steady speed called the **terminal velocity,** when:

air resistance = weight

Forces when skydiving.

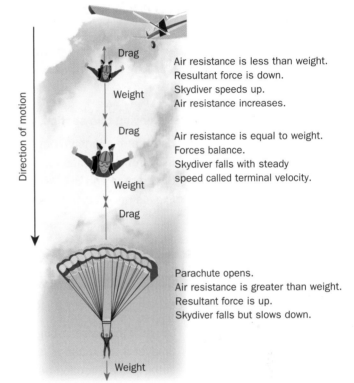

Direction of motion

Drag
Weight

Air resistance is less than weight.
Resultant force is down.
Skydiver speeds up.
Air resistance increases.

Drag
Weight

Air resistance is equal to weight.
Forces balance.
Skydiver falls with steady speed called terminal velocity.

Drag

Parachute opens.
Air resistance is greater than weight.
Resultant force is up.
Skydiver falls but slows down.

Weight

A falling object eventually reaches a terminal velocity.

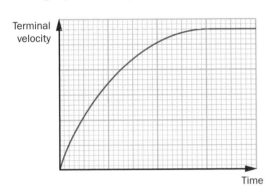

Terminal velocity

Time

In a vacuum, there is no air resistance so objects continue to fall with acceleration and acceleration due to gravity at 10 m/s.

> **It's a common mistake to think that there is always a force in the direction of movement.**

If an object is travelling at a steady speed there is no resultant force on it. When an object is slowing down the resultant force on it is opposite to its direction.

PROGRESS CHECK

1. What is the weight of 2 kg of sugar?
2. The driving force is 3000 N and friction is 900 N. What is the resultant force?
3. A car with steady speed of 100 km/h travels in a straight line. What is the resultant force?
4. Draw the forces on a book placed on a table.
5. How much does a 100 g apple weigh a) on Earth and b) on the Moon?
6. The driving force is 2500 N, friction is 800 N and air resistance is 1700 N.
   a) What is the resultant force?
   b) What happens to the car?

5.4 Acceleration and momentum

LEARNING SUMMARY

After studying this section, you should be able to:

- Use the relationship between force, mass and acceleration.
- Explain how momentum is calculated.
- Understand and use Newton's Second Law of Motion.
- Explain how collisions can be made safer.

Force, acceleration and momentum

AQA	P2	✓
OCR A	P4	✓
OCR B	P3	✓
EDEXCEL	P2	✓
WJEC	P2	✓
CCEA	P1	✓

A resultant force on an object causes it to accelerate. The acceleration is:

- Larger if the force is larger.
- Smaller if the mass is larger.
- In the same direction as the force.

> **KEY POINT**
>
> For a resultant force on an object:
>
> **force (N) = mass (kg) × acceleration (m/s^2)**
>
> $F = ma$
>
> where F = force, m = mass and a = acceleration.

The **momentum** of an object is:

- Larger for a larger velocity.
- Larger for a larger mass.
- In the same direction as the velocity.
- Measured in units called kg m/s or Ns (they are the same).

Momentum = mv Momentum = $-MV$

momentum (kg m/s or Ns) = mass (kg) × velocity (m/s)

momentum = mv where v = velocity

To remember what momentum depends on, think about 10 pin bowling. You are more likely to knock pins down with more momentum. This could be a ball with a lot of mass or a ball with a high velocity.

When a resultant force acts on an object, it causes a change in momentum in the same direction as the force (because the velocity changes). The bigger the force and the longer the time that the force acts, the bigger the change in momentum:

change in momentum (kg m/s or Ns) = force (N) × time (s)

Newton's second law of motion

AQA	P2	✓
OCR A	P4	✓
OCR B	P3	✓
EDEXCEL	P2	✓
WJEC	P2	✓
CCEA	P1	✓

KEY POINT

Newton's Second Law of Motion states:

When a resultant force acts on an object, it causes a change in momentum in the same direction as the force. The resultant force equals the rate of change of momentum.

$$\text{force (N)} = \frac{\text{change in momentum (kg m/s or Ns)}}{\text{time (s)}}$$

$$F = \frac{(mv - mu)}{t}$$

where t = time, u = initial velocity and v = final velocity.

The equation $F = ma$ is another way of saying this because acceleration

You do not need to be able to show or explain this.

$$a = \frac{(v - u)}{t}$$

$$F = \frac{m (v - u)}{t} = \frac{(mv - mu)}{t}$$

Safer collisions

AQA	P2	✓
OCR A	P4	✓
OCR B	P3, P5	✓
EDEXCEL	P2	✓
WJEC	P2	✓
CCEA	P1	✓

In a collision, a force brings your body to a sudden stop. The larger the stopping force on the body the more it is damaged. To reduce damage we must reduce the force:

change of momentum = force × time

For the same change in momentum, to reduce the force we must increase the time to stop. If the collision takes place over a longer time, say 0.5 s instead of 0.05 s – ten times as long – then the stopping force will only be one tenth of the size. The time of a collision can be increased by using:

- **Crumple zones** in the car. The front and back of the car are designed to crumple in a collision, increasing the distance and time over which the occupants are brought to a stop.

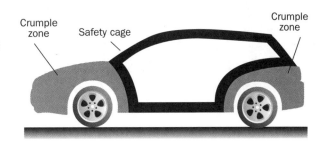

Crumple zone   Safety cage   Crumple zone

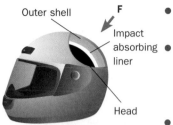

Outer shell

F

Impact absorbing liner

Head

> Do not confuse momentum with energy. Momentum has a direction and is measured in Ns or kg m/s. Energy, like mass, has no direction and it is measured in Joules (J). $1 J = 1 Nm$ – a different unit.

- The body hits the **airbag**, which is compressed, increasing the distance the body moves and the time it takes to stop.
- **Seatbelts** are designed to stretch slightly so that the body moves forward and comes to a stop more slowly with a smaller deceleration than it would if it hit the windscreen or front seats. After a collision the seatbelts should be replaced because having stretched once they may not work properly again.
- **Cycle and motor-cycle helmets** contain a layer of material which will compress on impact so that the skull is brought to a stop more slowly. They should be replaced after a collision as the material will be damaged and may not protect effectively again.

Other examples of reducing the force by increasing the time taken to stop include:

- Crash barriers that crumple on impact.
- Bending knees when landing after jumping.
- Bubble wrap.
- Sprung floors in gyms.

Objects that change shape absorb energy and sometimes heat up noticeably.

PROGRESS CHECK

1. Calculate the resultant force on a 1200 kg car accelerating at 3 m/s^2.
2. Calculate the momentum of a ball of mass 2 kg and velocity 5 m/s.
3. Calculate the momentum of a ball of mass 200 g and velocity 8 m/s travelling in the opposite direction.
4. A force of 50 N acts on a stationary object for 12 seconds. Calculate its gain in momentum.
5. A runaway truck of mass 1000 kg and velocity 12 m/s came to a sudden stop in 0.002 s.
   a) Calculate the stopping force on the truck.
   b) A crash barrier would have allowed the truck to stop over the greater time of 0.5 s. What would the stopping force have been?
   c) What difference would this make?

1. 1200 kg × 3 m/s^2 = 3600 N
2. 2 kg × 5 m/s = 10 kgm/s (or 10 Ns)
3. 0.2 kg × (−8 m/s) = − 1.6 kgm/s (or −1.6 Ns)
4. 50 N × 12s = 600 Ns
5. a) force = (1000 kg × 12 m/s) ÷ 0.002 s = 6 000 000 N
   b) force = (1000 kg × 12 m/s) ÷ 0.5 s = 24 000 N
   c) There would be less damage to the truck.

5.5 Action and reaction

LEARNING SUMMARY	After studying this section, you should be able to:
	- Explain that forces always occur in pairs.
	- Identify an interaction of pair of forces.
	- Use Newton's Third Law of Motion to calculate forces.
	- Use conservation of momentum to calculate velocities.

Interaction pairs

When two objects interact there is always an **interaction pair of forces**. As these skaters show, the man in red cannot push the man in blue without the man in blue pushing the man in red.

When two skaters push each other, they both move backwards.

In an interaction pair of forces, the two forces:

- Are always equal in size and opposite in direction.
- Always act on different objects.
- Are always the same type of force (for example, contact forces, gravitational forces, or magnetic forces).

> **KEY POINT**
>
> This idea is known as **Newton's Third Law of Motion** which states:
>
> *When two objects interact, the forces they exert on each other are equal and opposite and are called action and reaction forces.*

Friction and getting started

Friction is the reaction force needed for walking or wheeled transport.

Forces that make a wheel move forward.

Action Force   **Reaction Force**

To remember how friction gets you moving, think of trying to cycle or walk on a frictionless icy surface. There is no reaction force. You would slip backwards and never move forward.

- **Action force** is where the wheel pushes back on the road.
- **Reaction force** is where the road pushes forward on the wheel, which sends the wheel forward.

When you walk your foot pushes back on the ground and the ground pushes your foot forward.

Weight and the Earth

| OCR A | P4 | ✓ |
| OCR B | P3 | ✓ |

The weight of an object is the gravitational attraction towards the centre of the Earth. The other force of this interaction pair acts on the Earth. It is the gravitational attraction of the object attracting the whole Earth. We do not notice this effect because the mass of the Earth is so large.

An action and reaction pair of forces.

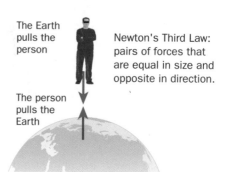

The Earth pulls the person

Newton's Third Law: pairs of forces that are equal in size and opposite in direction.

The person pulls the Earth

> Do not confuse interaction pairs of forces, which act on different objects, with balanced forces, which act on the same object.

The reaction force from a surface *balances* the weight of the object. It is not the reaction pair of the weight, because:

- It is a different type of force (a contact force, not gravitational).
- It acts on the same object.

Rockets and jet engines

OCR A	P4	✓
OCR B	P5	✓
WJEC	P3	✓
CCEA	P1	✓

Rockets and jet engines both produce hot exhaust gases. These are pushed out of the back of the engines. There is an equal and opposite reaction force that sends the rocket or jet forward.

Conservation of momentum and rocket propulsion.

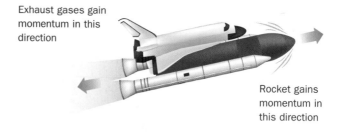

Exhaust gases gain momentum in this direction

Rocket gains momentum in this direction

> For OCR B you need to be able to apply particle theory to rocket exhaust gases.

Conservation of momentum

AQA	P2	✓
OCR A	P4	✓
OCR B	P5	✓
EDEXCEL	P2, P3	✓
WJEC	P3	✓
CCEA	P1	✓

When two objects **collide** or **explode apart** there is an equal and opposite force on each object, and they interact (push against each other) for the same time. This means that the change in the momentum of the objects is equal and opposite. Another way to say this is that **momentum is conserved**.

The total momentum of the two objects before the collision or explosion is the same as the total momentum after.

Example: When two objects collide and stick together.

Two objects colliding and sticking together.

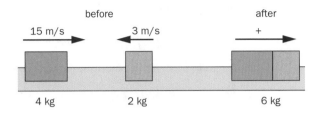

Before the collision momentum:

= 4 kg × 15 m/s + 2 kg × (–3) m/s = (60 – 6) kg m/s

After the collision:

momentum = 6 kg × v

So, because of conservation of momentum

v = 9 m/s

Recoil

OCR A	P4	✓
OCR B	P5	✓
WJEC	P3	✓
CCEA	P1	✓

When a bullet leaves a gun, action and reaction, or **conservation of momentum**, tell us the gun must recoil.

Example: A 0.8 g paintball is fired at 80 m/s from a 3 kg paintball marker.

Firing a paintball.

$(0.8 \times 80) + (3000g \times x) = 0$

$64 + 3000x = 0$

$x = \dfrac{-64}{3000}$

$x = 0.02$

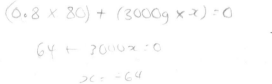

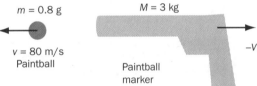

m = 0.8 g

v = 80 m/s
Paintball

M = 3 kg

–V?

Paintball
marker

Conservation of momentum: $m\,v = M\,V$

Recoil velocity:

$$V = \frac{0.8 \text{ g} \times 80 \text{ m/s}}{3000 \text{ g}} = 0.02 \text{ m/s}$$

PROGRESS CHECK

1. A girl pushes on a wall with a force of 5 N. Describe the reaction force.
2. Why is it difficult to walk on ice?
3. How does a rocket move in outer space where there is nothing to push against to get moving?
4. When you release a partly inflated balloon it flies around as it deflates. Explain why.
5. A book is placed on a table. What are the two interaction pairs of forces?
6. A toy car with mass 0.5 kg and speed 4 m/s collides with a toy truck of mass 2 kg. They both stop. What was the speed of the truck?

5.6 Work and energy

LEARNING SUMMARY

After studying this section, you should be able to:

- Relate energy transfers to work done.
- Calculate changes in gravitational potential energy.
- Calculate the kinetic energy of moving objects.
- Recognise when the change in GPE = the change in KE and use this in calculations.

Work and energy

AQA	P2	✓
OCR A	P4	✓
OCR B	P3	✓
EDEXCEL	P2	✓
WJEC	P2	✓
CCEA	P1	✓

When a **force** makes something move, **work** is done. The amount of work done is equal to the amount of energy transferred. Work and energy are measured in joules (J):

> **KEY POINT**
>
> **work done by a force (J) = force (N) × distance moved by force in direction of the force (m)**
>
> When work is done *by* something it loses energy, when work is done *on* something it gains energy.

Gravitational potential energy

AQA	P2	✓
OCR A	P4	✓
OCR B	P3	✓
EDEXCEL	P2	✓
WJEC	P2	✓
CCEA	P1	✓

A rollercoaster at the top of a slope has stored energy, which is called **gravitational potential energy (GPE)**, or sometimes **potential energy (PE)**. This is the stored energy that an object has because of its position, in this case, higher above the surface of the Earth.

Example: Doing work – increasing the GPE.

Doing work – increasing the GPE.

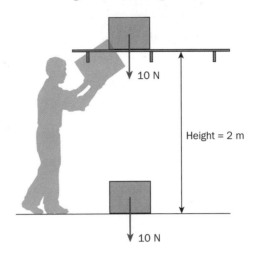

10 N

Height = 2 m

10 N

> It's the change in GPE that depends on the change in height.

When you lift a 10 N weight (a mass of 1 kg) from the floor to a high shelf, a height difference of 2 m, you have done work on the weight.

The work done = 10 N × 2 m = 20 J and this is equal to the increase in the GPE of the weight.

> **KEY POINT**
>
> **Change in GPE (J) = weight (N) × vertical height difference (m)**

> **KEY POINT**
>
> **Change in GPE = *m g h*;** where *g* = the gravitational field strength (N/kg) *m* = mass (kg) and *h* = height change (m).

Example: To calculate a change in height from change in GPE.

A mass of 2 kg is lifted up and gains 400 J of GPE (gravitational field strength, *g* = 10N/kg).

Change in height *h*:

$$= \frac{400 \text{ J}}{2 \text{ kg} \times 10 \text{ N/kg}} = 20 \text{ m}$$

Kinetic energy

AQA	P2	✓
OCR A	P4	✓
OCR B	P3	✓
EDEXCEL	P2	✓
WJEC	P2	✓
CCEA	P1	✓

An object that is moving has **kinetic energy (KE)**. The energy depends on the mass of the object and on the square of the speed. Doubling the speed gives four times the energy.

Example: An air hockey puck, floating on an air table, is almost frictionless. A force does work on the puck – it pushes it a small distance. Energy is transferred and the kinetic energy of the puck increases – it speeds up. When the force stops the puck moves at a constant speed across the table – its kinetic energy is now constant.

> **KEY POINT**
>
> kinetic energy (J) = $\frac{1}{2}$ × mass (kg) × [speed (m/s)]^2

Energy does not have a direction. Speed or velocity can be used to calculate the kinetic energy.

KE = $\frac{1}{2} m v^2$ where m = mass (kg) v = speed (m/s).

Example: To calculate the speed of an object from its KE.

A mass of 2 kg has 400 J of KE.

speed $v = \sqrt{\dfrac{(2 \times 400\,J)}{2\,kg}}$ = 20 m/s

Calculations with GPE and KE

AQA	P2	✓
OCR A	P4	✓
OCR B	P3	✓
EDEXCEL	P2	✓
WJEC	P2, P3	✓
CCEA	P1	✓

When frictional forces are small enough to be ignored the transfer of energy between KE and GPE can be used to calculate heights and speeds.

Example: A car is driven by a trackside motor to the top of a rollercoaster and then freewheels down the slope. Assume that the GPE at the bottom of the slope is zero.

Transferring energy from GPE to KE and back again.

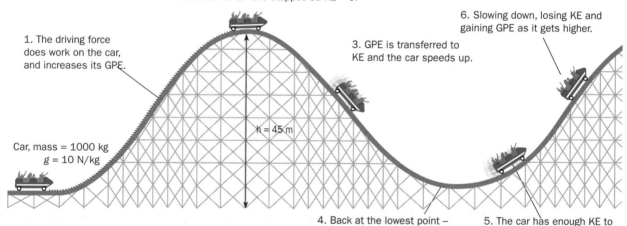

2. The highest point – the car has maximum GPE. It is stopped so KE = 0.

1. The driving force does work on the car, and increases its GPE.

3. GPE is transferred to KE and the car speeds up.

6. Slowing down, losing KE and gaining GPE as it gets higher.

h = 45 m

Car, mass = 1000 kg
g = 10 N/kg

4. Back at the lowest point – maximum KE and GPE = 0.

5. The car has enough KE to continue up the next slope.

Increase in GPE of car:

= $m g h$ = 1000 kg × 10 N/kg × 45 m = 450 000 J

Assuming there are no friction forces as the car travels down the slope:

Loss of GPE = gain in KE

gain in KE = 450 000J

450 000 J = $\frac{1}{2} m v^2$ = $\frac{1}{2}$ × 1000 kg × v^2

v^2 = 900 (m/s)^2 so speed v = 30 m/s

> **Remember to square a number by multiplying it by itself. A calculator is useful for finding the square root of a number.**

5.7 Energy and power

LEARNING SUMMARY

After studying this section, you should be able to:

- Explain and use the principle of conservation of energy.
- Explain what happens to the GPE of an object falling with terminal velocity.
- Calculate shopping distances.
- Describe factors that affect stopping distances.
- Use the relationship between power, energy transferred and home.

Conservation of energy

AQA	P1	✓
OCR A	P4	✓
EDEXCEL	P1, P2	✓
WJEC	P2	✓
CCEA	P1	✓

The **principle of conservation of energy** says that the total energy always remains the same. When energy is transferred to the surroundings by heating due to frictional forces it is no longer useful, but it is not lost. We say it has been **dissipated** (spread out) as heat.

The relationship 'gain in KE = loss in GPE' is only true for a falling object if the air resistance (or drag) is small and can be ignored, or if the object is falling in a vacuum.

Work against friction

AQA	P2	✓
OCR A	P4	✓
OCR B	P3	✓
WJEC	P2	✓
CCEA	P1	✓

A skydiver eventually reaches **terminal velocity**. She is still falling, so GPE is being lost, but no KE is being gained. The energy is being used to do work against the frictional force (air resistance.) The skydiver and surrounding air will heat up.

Forces when skydiving.

Air resistance

Weight

The space shuttle, with a lot of KE, needed heat proof tiles to protect it from the heat resulting from doing work against air resistance when it re-entered the Earth's atmosphere.

When a cyclist pedals, but travels at a steady speed, work is done against air resistance and friction. Energy is transferred and heats the bicycle and surroundings. No energy is being transferred as KE to the bicycle unless it speeds up.

Stopping distances

AQA	P2	✓
OCR B	P3	✓
EDEXCEL	P2	✓
WJEC	P2	✓

The distance travelled between the driver noticing a hazard and the vehicle being stationary is called the **stopping distance**.

> **KEY POINT**
>
> - Stopping distance = thinking distance + braking distance.
> - **Thinking distance** is distance travelled during the driver's reaction time – the time between seeing the hazard and applying the brakes.
> - **Braking distance** is the distance travelled while the vehicle is braking.

This diagram shows the shortest stopping distances at different speeds

The stopping distance increases with speed.

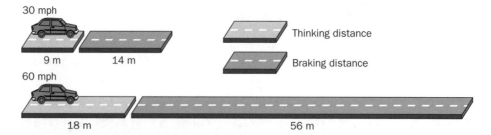

30 mph

9 m   14 m

Thinking distance

Braking distance

60 mph

18 m   56 m

When speed doubles:

- Thinking distance doubles.
- Braking distance is four times as far.

The stopping distances are also longer if:

- The driver is tired, affected by some drugs (including alcohol and some medicines) or distracted and not concentrating, so thinking distance is increased.
- The road is wet or icy or the tyres or brakes are in poor condition. The friction forces will be less so the braking distance is increased.
- The vehicle is fully loaded with passengers or goods. The extra mass reduces the deceleration during braking, so the braking distance is increased.

These stopping distances are taken into account when setting road speed limits. Drivers should not drive closer than the thinking distance to the car in front, to allow for time to react. They should reduce speed in bad weather to allow for the increased braking distance.

Braking and kinetic energy

AQA	P2	✓
OCR B	P3	✓
EDEXCEL	P2	✓
WJEC	P2	✓

- When speed doubles reaction time is the same:

  thinking distance = speed × reaction time (thinking distance doubles)

- Work done by the brakes against friction = loss in KE
  braking force × braking distance = $\frac{1}{2} mv^2$

  $$\text{braking distance} = \frac{\text{mass} \times \text{speed}^2}{2 \times \text{braking force}}$$

  The braking distance is four times as far.

> For OCR B you need to be able to draw and interpret graphs of stopping distance and about ABS brakes.

KEY POINT

The **thinking distance** depends on **speed**.

The **braking distance** depends on **(speed)^2**.

Example: At three times the speed, braking distance is nine times as far.

Power

AQA	P2	✓
OCR B	P3	✓
EDEXCEL	P2	✓
CCEA	P1	✓

KEY POINT

Power is the work done, or energy transferred, divided by time. Power is measured in watts (W).

$$\text{power (W)} = \frac{\text{work done or energy transferred (J)}}{\text{time (s)}}$$

Power is also the rate of energy transfer.

> Units can help you to remember how things are related and do calculations. A watt is a joule per second, so divide energy (joules) by time (seconds) to get power (watts).

Example: A 7.5 kW crane lifts a 3000 N weight up a height of 10 m. How long does it take?

$$7.5 \text{ kW} = \frac{3000 \text{ N} \times 10 \text{ m}}{t}$$

$$t = \frac{30\,000 \text{ J}}{7500 \text{ W}} = 4 \text{ s}$$

> Don't forget to use the correct units in your calculations. It can help to change everything to the basic units, for example kilowatts to watts. An answer without units is not an answer – it's just a number.

PROGRESS CHECK

1. Under what conditions is the 'loss in GPE = gain in KE' ?
2. What happens to the GPE when an object falls with terminal velocity?
3. Give an example where stopping distances would be longer than shown in the diagram.
4. A worker does 1000 J of work in 5 s. What power was used?
5. What happens to the energy transferred by a pedalling cyclist when travelling at a steady speed?
6. Calculate the stopping distance of the car in the diagram when travelling at 90 mph.

1. When the frictional forces are so small they can be ignored, or in a vacuum.
2. The GPE is transferred to heating the air and the object.
3. Any of listed examples e.g. driver has been drinking or poor condition of tyres.
4. 1000 J ÷ 5 s = 200 W
5. The cyclist is doing work against friction forces the energy is transferred as heat to the bicycle, road and air.
6. 3 × 30mph so thinking distance = 3 × 9 m braking distance = 9 × 14 m stopping distance = 153 m

Sample GCSE questions

1 An aircraft with a mass of 9000 kg is flying at a level height above the ground.

Lift 90 kN

Thrust 100 kN ← → Drag 5 kN

Weight (gravity) 90 kN

(a) Calculate the acceleration of the aircraft. **[4]**

$force = 100 \text{ kN} - 5 \text{ kN} = 95 \text{ kN}$

$F = ma$

$95 \text{ kN} = 9000 \text{kg} \times a$

$a = \dfrac{95 \text{ kN}}{9000 \text{ kg}}$

$a = 10.6 \text{ m/s}^2$

Work out the resultant force forwards

Show which equation you are using

Write down the calculation – if you do it incorrectly you may still get a mark.

Remember the units

(b) Draw a ring around the two quantities of the aircraft that are increasing. **[2]**

acceleration

gravitational potential energy

height

(kinetic energy)

(momentum)

(c) After some time, the aircraft stops accelerating and travels at a steady speed although the driving force is still 100 kN. Describe how the speed of the aircraft changes during its level flight and explain why this happens. *The quality of your written communication will be assessed in this answer.* **[6]**

As the aircraft accelerates it gets faster and the drag force on it increases. This is because the drag force increases with speed. As the drag force increases the resultant force forwards is reduced. This means that the acceleration is reduced. So as the aircraft gets faster its acceleration gets lower until eventually, at the top speed, the acceleration is zero and the aircraft no longer accelerates. This speed is called the terminal velocity.

[Total = 12]

Marks will be awarded depending on the number of relevant points included in the answer and the spelling, punctuation and grammar. In this question there are 6 or 7 relevant points so 5 or 6 points with good spelling punctuation and grammar will gain full marks.

Sample GCSE questions

2 This graph shows the velocity of a car for part of its journey.

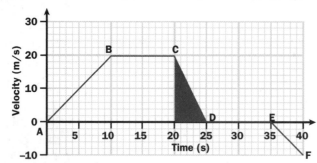

(a) Describe the motion of the car during the journey. *The quality of your written communication will be assessed in this answer.* **[6]**

At point A the car is stationary. It accelerates with constant acceleration for 10 seconds to a speed of 20 m/s at point B. From B to C it travels at a steady speed of 20 m/s. From C to D it slows down to a stop over 5 seconds. From D to E it is stationary for 10 seconds and from E to F it accelerates in the opposite direction for 5 seconds, reaching a speed of 10 m/s back towards the start.

(b) What is the acceleration of the car between A and B?

$$acceleration = \frac{change\ in\ speed}{time}$$

Change in speed = (20 m/s − 0 m/s) time = 10 s
Acceleration = 20 m/s ÷ 10 s = 2 m/s²

Acceleration = ...2......m/s² **[2]**

← State the equation you use.

The acceleration is the gradient of the line, so, instead of using the equation, you could have explained that you were working out the gradient.

(c) What is the distance travelled between points C and D?

distance = shaded area under the line
Area of triangle made by CD = ½ × (20 m/s × 5 s) = 50 m ←

Distance travelled = ...50.......m **[2]**

Show what you are calculating, you may still get a mark if you later make a mistake.

(d) The distance travelled between the point A and point C is 300 m.

(i) Calculate the average velocity between point A and point C.

$$v = \frac{total\ distance}{time} \qquad v = \frac{300\ m}{20\ s} = 15\ m/s$$

(ii) Compare your value for the average velocity with the velocity at point C on the graph, and explain why they are different. **[3]**

The velocity at point C on the graph is 20 m/s which is greater than the average velocity between A and C of 15 m/s. This is because 20 m/s is the instantaneous velocity and not the average velocity. The average velocity is reduced by the fact that between A and B the velocity was less than 20 m/s.

[Total = 13]

Exam practice questions

1 These statements describe what happens during a parachute jump.

A	The parachute opens – increasing the drag force
B	His speed increases and the drag force increases
C	The drag force becomes equal to his weight
D	The skydiver steps from the aircraft and falls, he accelerates at 10 m/s^2
E	He falls with a constant speed
F	His speed decreases

Put them in the correct order, the first one has been done for you. **[4]**

D					

2 In a test a 1200 kg car crashes into the back of a 3000 kg lorry:

v = 15m/s

v = 20m/s

M = 1200kg   M = 3000kg

The car has velocity 20 m/s.

(a) Calculate the momentum of the car before the collision. **[2]**

....

(b) In a collision the total momentum of the car and the lorry **[1]**

    A is increased

    B is reduced

    C stays the same

    D may do any of the above.

    ☐

(c) The momentum of the lorry is 45 000 kgm/s.
The car and truck stick together after the collision.
Calculate the speed of the car and truck after the collision. **[6]**

....

....

....

....

....

[Total = 9]

Exam practice questions

3 Roller coaster track:

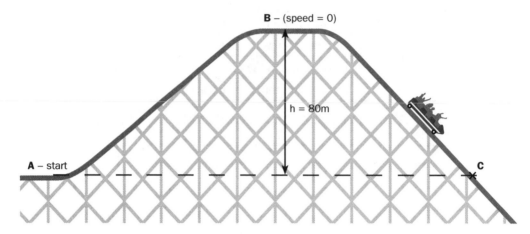

The mass of a car is 800 kg

Gravitational field strength = 10 N/kg

Assume there are no energy losses due to friction as the car goes round the track.

(a) Calculate the increase in potential energy when the car moves from A to B. **[2]**

....

....

(b) What is the kinetic energy at point C? Explain how you know this. **[2]**

....

....

(c) Calculate the speed of the car at point C. **[2]**

....

....

(d) At the end of the ride, the kinetic energy of the car is 25 000 J. It brakes to a stop in 10 m. What is the braking force? **[2]**

....

....

[Total = 8]

4 This van has a mass of 2000 kg.

Driving force ←   → Resistive force 5000N

Exam practice questions

When the van travels at a steady speed of 30 m/s the resistive force is 5000 N.

(a) What is the driving force on the van? **[1]**

....

(b) Calculate the kinetic energy of the van. **[2]**

....

....

[Total = 3]

5 This table shows the shortest stopping distances for a car at different speeds.

Speed mph	Thinking distance (m)	Braking distance (m)	Stopping distance (m)
20	6	6	12
40	12	24	
60	18	55	73

(a) Work out the missing value for the stopping distance at 40 mph and write it in the table. **[1]**

(b) These are the shortest stopping distances. Describe some factors that will increase the stopping distance and explain whether each factor increases the thinking distance, or the braking distance. *The quality of your written communication will be assessed in this answer.* **[6]**

....

....

....

....

....

....

(c) In a collision test when a dummy hits the windscreen it stops in 0.001 s.

When the dummy is wearing a seatbelt it stops in 0.02 s.

(i) What happens to the seatbelt to make the stopping time longer? **[1]**

....

(ii) Explain why the longer stopping time reduces the damage to the dummy. **[2]**

....

....

[Total = 10]

6 Electricity

The following topics are covered in this chapter:

- **Electrostatic effects**
- **Uses of electrostatics**
- **Electric circuits**
- **Voltage or potential difference**
- **Resistance and resistors**
- **Special resistors**
- **The mains supply**

6.1 Electrostatic effects

LEARNING SUMMARY

After studying this section, you should be able to:

- Recall that electric charge can be positive or negative.
- Describe how objects can become charged.
- Describe forces between charged objects.
- Explain how objects are earthed.
- Explain some dangers and annoying effects of electric charge.

Electric charge

AQA	P2	✓
OCR A	P5	✓
OCR B	P4	✓
EDEXCEL	P2	✓
CCEA	P2	✓

Electric charge can be **positive** or **negative**. **Electrons** are particles with a negative electric charge. They can move freely through a **conductor**, for example any type of metal, but cannot move through an **insulator**.

Two objects attract each other if one is positively charged and the other is negatively charged. Two objects with similar charge (both positive or negative) repel.

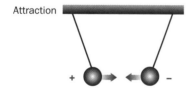
Attraction

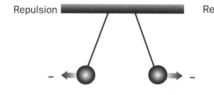
Repulsion

Repulsion

Electrostatic effects are caused by the transfer of electrons. (It is also sometimes called static electricity.) When insulators are rubbed, electrons are rubbed off one material and transferred to the other.

A polythene rod rubbed with a duster picks up electrons from the duster and becomes negatively charged, leaving the duster positively charged.

The insulated rod and the cloth have opposite charge.

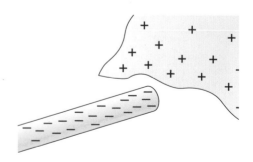

> **KEY POINT**
>
> Like charges repel. Unlike charges attract.

A Perspex rod rubbed with a duster loses electrons to the duster and becomes positively charged, leaving the duster negatively charged.

Materials that are positively charged have missing electrons. Materials that are negatively charged have extra electrons.

Charging and discharging

AQA	P2	✓
OCR A	P5	✓
OCR B	P4	✓
EDEXCEL	P2	✓
CCEA	P2	✓

Conductors cannot be charged unless they are completely surrounded by insulating materials, such as dry air and plastic, otherwise the electrons flow to or from the conductor to discharge it.

An **insulated conductor** can be charged by rubbing it with a duster, or touching it with a charged rod. Some electrons are transferred, so that the charge is spread out over both objects.

A conductor can be discharged by touching it with another conductor, for example a wire, so that electrons can flow along the wire and cancel out the charge.

> Remember that it is the electrons that move from one object to another.

To stop conductors becoming charged they can be **earthed**. A thick metal wire is used to connect them to a large metal plate in the ground. This acts as a large reservoir of electrons. Electrons flow so quickly to, or from, earth that objects connected to earth do not become charged.

Dangerous or annoying?

AQA	P2	✓
OCR B	P4	✓
EDEXCEL	P2	✓
CCEA	P2	✓

The human body conducts electricity. When a large flow of charge affects our nerves and muscles we call this an **electric shock**.

- Small electrostatic shocks are not harmful.
- Larger shocks can be dangerous to people with heart problems because a flow of charge through the body can stop the heart.
- Lightning is a very large electrostatic discharge. When it flows through a body it is often fatal.

Charged objects, like plastic cases and TV monitors, attract small particles of dust and dirt. Clothing can be charged as you move and 'clings' to other items of clothing, or the body. Synthetic fibres are affected more than natural fibres as they are better insulators. On a dry day charge can build up on you. When you touch metal, for example a car door, the charge flows from you to the metal, and you get a shock. A **spark** occurs when electrons jump across a gap. This can cause an explosion if there are:

Inflammable vapours like petrol or methanol

Inflammable gases like hydrogen or methane

Powders in the air, like flour or custard, which contain lots of oxygen – as a dust they can explode

Discharging safely

AQA	P2	✓
OCR B	P4	✓
EDEXCEL	P2	✓
CCEA	P2	✓

Lorries containing flammable gases, liquids and powders are connected to earth before loading or unloading. Aircraft are earthed before being refuelled. This prevents charge from building up on metal pipes or tanks when the loads are moved, so there is no danger of a spark igniting the load.

Anti-static sprays, liquids and cloths stop the buildup of static charge. These work by increasing the amount of conduction, sometimes by attracting moisture because water conducts electricity.

If you stand on an insulating mat, or wear shoes with insulating soles, when you touch a charged object this will reduce the chance of an electric shock because the charge will not flow through you to earth. You become charged and stay charged until you touch a conductor.

PROGRESS CHECK

1. Why is a plastic rod attracted to a cloth it has been rubbed with?
2. What particles are transferred when a balloon is rubbed with a cloth?
3. How many types of electric charge are there?
4. Why do you become charged walking on a nylon carpet, but not on a woollen carpet?
5. When a plastic rod and a Perspex rod are both charged by rubbing they attract each other.
   a) What does this tell you about the charges on them?
   b) Would you expect two rubbed polythene rods to attract or repel?
6. Why are aircraft earthed before being refuelled?

1. They have opposite charges, which attract.
2. Electrons
3. Two – positive and negative
4. Nylon is a better insulator. Some of the charges are conducted away through the wool carpet.
5. a) They are opposite b) They repel because they are the same, and would get the same charge.
6. So that there is not a build up of charge which could cause a spark. A spark would ignite the fuel.

6.2 Uses of electrostatics

LEARNING SUMMARY	**After studying this section, you should be able to:**
	• Describe and explain some uses of electrostatics, including:
	  – electrostatic precipitators
	  – electrostatic paint spraying
	  – defibrillators.

Electrostatic precipitators

OCR B   P4   ✓

Electrostatic precipitators remove dust or smoke particles from chimneys, so that they are not carried out of the chimney by the hot air.

An electrostatic precipitator.

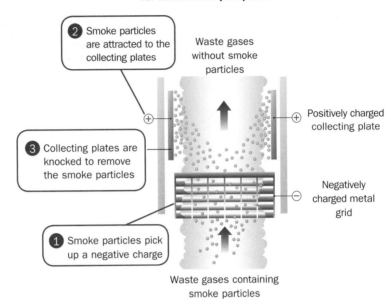

2 Smoke particles are attracted to the collecting plates

Waste gases without smoke particles

3 Collecting plates are knocked to remove the smoke particles

Positively charged collecting plate

Negatively charged metal grid

1 Smoke particles pick up a negative charge

Waste gases containing smoke particles

- Charged metal grids are put in the chimneys.
- The smoke particles pass through the grids and become charged.
- Plates at the side are oppositely charged to the grids.
- The smoke particles are attracted and stick to the plates.
- The smoke particles clump together on the plates to form larger particles.
- The plates are knocked and the large particles fall back down the chimney into containers.

> When you are describing how these applications work, explain what happens to the electrons in each case and how this affects the charge on the objects.

The grids are connected to a high voltage. They attract or repel charges in the smoke particles, so the particles become charged. The grids are positively charged in some designs and negatively charged in others. If the grids are positively charged, the plates are earthed. The smoke or dust particles lose electrons and become positively charged. They induce a negative charge on the earthed metal plate and are attracted to the plate. If the grids are negatively charged the plates are positively charged. The smoke or dust particles gain electrons and become negatively charged. They are attracted to the positively charged metal plates.

Spray painting

OCR B   P4   ✓
EDEXCEL P2   ✓

Before spray painting, the paint and the object are given different charges so that the paint is attracted to the object.

- The spray gun is charged so that it charges the paint particles.
- The paint particles repel each other to give a fine spray.
- The object is charged with the opposite charge to the paint.
- The object attracts the paint.
- The paint makes an even coat, it even gets underneath and into parts that are in shadow.
- Less paint is wasted.

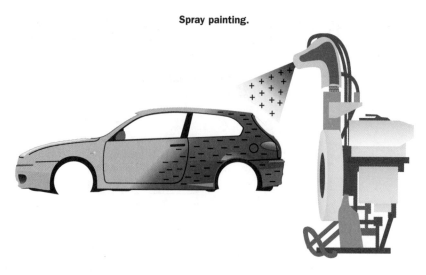

Spray painting.

In the example shown in the diagram, the paint droplets have lost electrons, so they are positively charged. The metal door panel is either connected to earth, or to a negative terminal so it is negatively charged. The positively charged drops are attracted to the negatively charged door. The door does not lose its negative charge, so it continues to attract positive droplets. As the coat of paint builds up, the new droplets will be attracted to the parts of the door that are the most negative – those in the shadows and cracks that the paint would not usually reach.

Crop spraying

| OCR B | P4 | ✓ |
| EDEXCEL | P2 | ✓ |

Crop spraying.

Fertiliser and insecticide spray nozzles are charged so that the droplets leaving the nozzle are charged. They repel each other and they are attracted to uncharged objects like the plants. The fine droplets cover the plant better and do not collect into large drops. They are less likely to drift in the wind and get wasted. This means that much less is used, which saves money and is better for the environment.

Defibrillators

OCR B   P4   ✓

When the heart beats, the heart muscle contracts. A **defibrillator** is used to start the heart when it has stopped.

The dotted lines show the path of the charge through chest, and heart.

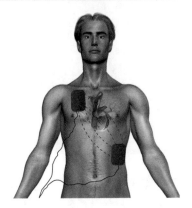

- Two electrodes called paddles are placed on the patient's chest.
- The paddles must make a good electrical contact with the patient's chest.
- Everyone including the operator must 'stand clear' so they don't get an electric shock.
- The paddles are charged.
- The charge is passed from one paddle, through the chest to the other paddle to make the heart muscle contract.

The paddles take a few moments to charge up and then the discharge happens quickly. The electrons move through the heart muscle. Often the heart has not stopped, but has lost its steady rhythm. The defibrillator allows the heart to restart beating to its normal rhythm again.

> **Remember these uses:**
> **Smoke, Sprays, and Shock.**

PROGRESS CHECK

1. Why do the plates in the electrostatic precipitator have the opposite charge to the grids?
2. What would happen if the paint drops had the same charge as the car body?
3. Give an advantage of electrostatic crop spraying.
4. Why is it important to make sure no one except the patient gets a shock from a defibrillator?
5. Why are the grids in an electrostatic precipitator connected to a high voltage?
6. What would happen if the car door was not connected to earth or a negative charge?

1. The grids will attract (or repel) electrons from the smoke particles, which will be left with the same charge as the grid, so the plates need to have the opposite charge to attract the charged smoke particles.
2. They would be repelled and would not stick to the car body.
3. Better coverage or less wastage.
4. The shock could stop their heart.
5. So that they will charge the smoke particles that come close to them, by attracting or repelling electrons.
6. The positive drops would give the car body a positive charge so that eventually the car body would be positively charged and the paint would be repelled.

6.3 Electric circuits

LEARNING SUMMARY

After studying this section, you should be able to:

- Recognise circuit symbols.
- Draw electric circuits.
- Describe how to use an ammeter.
- Explain how charge flows in series and parallel circuits.
- Calculate electric charge and currents in circuits.

Circuit symbols

AQA	P2	✓
OCR A	P5	✓
OCR B	P6	✓
EDEXCEL	P2	✓
WJEC	P2	✓
CCEA	P2	✓

To learn the symbols, draw or print them onto cards – one card with the word and one card with the symbol. Then use them to play 'pairs.' Place them face down and turn two over at a time. If you get a 'pair' of the matching word and symbol you keep it. If not you place them face down again. The winner is the person with the most pairs.

Take care when drawing circuit diagrams. Although the shape of the connecting wires does not matter they must join the components properly – electricity can't flow through gaps. The ammeter and voltmeter symbols are circles, not squares, and the symbol is 'A' not 'a'.

Component	Symbol	Component	Symbol
switch (open)		lamp	
switch (closed)		fuse	
cell		fixed resistor	
battery		variable resistor	
ammeter		light dependent resistor (LDR)	
voltmeter		thermistor	
junction of conductors		diode	
motor		generator	
power supply		a.c. supply	

Electric current

AQA	P2	✓
OCR A	P5	✓
OCR B	P4, P6	✓
EDEXCEL	P1, P2	✓
WJEC	P2	✓
CCEA	P2	✓

Electric current:

- is a flow of **electric charge**.
- only flows if there is a compete circuit. Any break in the circuit switches it off.
- is measured in **amps** (A) using an **ammeter**.
- is not used up in a circuit. If there is only one route around a circuit the current will be the same wherever it is measured.
- transfers energy to the components in the circuit.

A **series circuit** is a circuit with only one route around it. The current measured on each ammeter will be the same.

A series circuit.

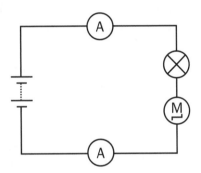

A **parallel circuit** has more than one path for the current around the circuit. In this circuit there are two paths, marked in red and blue, around the circuit. The current measured on ammeters B and C adds up to give the current measured on ammeter A and on ammeter D.

A parallel circuit.

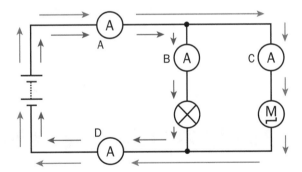

Electric current is a flow of **positive charge** so the direction of the current is opposite to the direction of the electron flow, because electrons are negatively charged.

> **KEY POINT**
>
> Electric charge is measured in **coulombs** (C). The amount of electric charge passing a point in the circuit depends on the current:
>
> **Charge (C) = current (A) × time (s)**
>
> $Q = I\,t$

Metal conductors contain lots of electrons that are free to move. When the battery makes the electrons move, they flow in a continuous loop around the circuit. In insulators there are few charges that are free to move.

KEY POINT

Batteries supply direct current, (d.c.) so the charges always move in the same direction from the positive terminal, around the circuit to the negative terminal. Mains electricity is produced by generators and the charges reverse direction. This is called alternating current (a.c.)

At a junction in a circuit, the total current flowing into the junction must be the same as the total current flowing out of the junction.

PROGRESS CHECK

1. In the series circuit with two ammeters on page 125, if they both read 0.2 A what would be the reading on a third ammeter placed between the lamp and the motor?
2. In the parallel circuit on page 125, if ammeter B reads 0.3 A and ammeter C reads 0.5 A what is the reading on:
   a) Ammeter A?
   b) Ammeter D?
3. In the parallel circuit on page 125, if ammeter B reads 500 mA and ammeter A reads 900 mA what is the reading on:
   a) Ammeter C?
   b) Ammeter D?
4. If a current of 2 A is switched on for 10 s, how much charge has flowed?
5. In the parallel circuit on page 125, what will always be true about the readings on ammeters A, B and C?

5. reading A = reading B + reading C
4. 20 C
3. a) 400 mA b) 900 mA
2. a) 0.8 A b) 0.8 A
1. 0.2 A

6.4 Voltage or potential difference

After studying this section, you should be able to:

- Describe how to use a voltmeter.
- Explain and use the term potential difference.
- Use the relationships between potential difference, work and charge.
- Calculate voltages in series and parallel circuits.

Voltage or potential difference

AQA	P2	✓
OCR A	P5	✓
OCR B	P6	✓
EDEXCEL	P1, P2	✓
WJEC	P2	✓
CCEA	P2	✓

Students often confuse voltmeters and ammeters. Always say 'voltage across' and 'current through'.

This will remind you that to measure the current flowing through a component you must connect the ammeter in line, so that the current flows through it. To measure the voltage across the component you must connect the voltmeter across the component making a connection on either side of it.

KEY POINT

Voltage is also called **potential difference (p.d)**. Voltage is:

- Measured between two points in a circuit.
- Measured in **volts** (V) using a **voltmeter**.

The higher the voltage of a battery the greater the 'push' on the charges in the circuit.

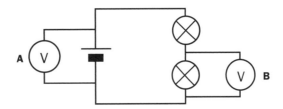

This diagram shows how to connect voltmeter A to measure the voltage supplied by the battery, and how to connect voltmeter B to measure the voltage across one of the lamps.

Potential difference, or voltage, is a measure of energy transferred to (or from) the charge moving between the two points.

In the diagram:

- Voltmeter A is measuring the energy transferred *to* the charge.
- Voltmeter B is measuring the energy transferred *from* the charge.

Remember that 'a volt is a joule per coulomb.' Add to this that 'a coulomb is an amp second' and you can work out most of the electricity relationships you need.

KEY POINT

The potential difference (voltage) between two points is the work done (energy transferred) per coulomb of charge that passes between the two points.

$$\text{Potential difference (V)} = \frac{\text{Work done (J)}}{\text{Charge (C)}}$$

$$V = \frac{W}{Q}$$

Voltage in series or parallel

AQA	P2	✓
OCR A	P5	✓
OCR B	P6	✓
EDEXCEL	P2	✓
WJEC	P2	✓
CCEA	P2	✓

When components are connected in **series** the voltage, or p.d., of the power supply is shared between the components.

Adding the measurements on the three voltmeters gives the power supply p.d.

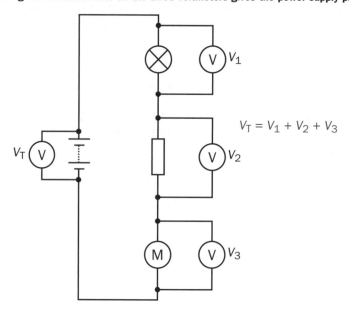

$$V_T = V_1 + V_2 + V_3$$

The measurements on all the voltmeters are the same.

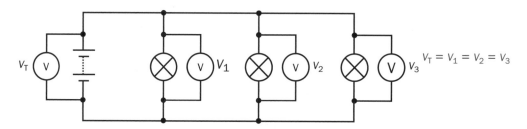

$$V_T = V_1 = V_2 = V_3$$

When components are connected in **parallel** to a power supply, the voltage, or p.d., across each component is the same as that of the power supply.

Connecting cells together

AQA	P2	✓
OCR A	P5	✓
OCR B	P6	✓
EDEXCEL	P2	✓

Only identical cells should be connected together. In series the p.d. will be the sum of the p.d.s of the cells. In parallel, the p.d. is unchanged. The current will be larger when the cells are in series. In parallel, the current is unchanged, but the cells will last longer because there is more stored charge.

Cells connected in series and in parallel.

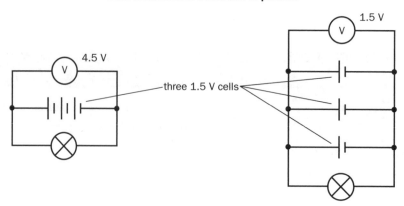

1. A motor, a resistor and a lamp are connected in series. The battery voltage is 12 V, the voltage across the motor is 6 V and across the resistor is 4 V.
   a) What is the voltage across the lamp?
   b) Does this mean that the current will be different in each component? Explain your answer.
2. In the parallel circuit on page 128 (middle), if the battery voltage is 9 V:
   a) what is the voltage across each of the lamps?
   b) Does this mean that the lamps will be equally bright? Explain your answer.
3. If the voltage across a lamp is 9 V and 10 C of charge flows through the lamp how much energy has been transferred?
4. What voltage would be supplied by five 1.5 V cells:
   a) in series?
   b) in parallel?
5. What advantage is there to connecting the five cells in parallel?

5. They will last longer – there is five times as much stored charge.
4. a) 5 × 1.5 V = 7.5 V b) 1.5 V
3. 9 V × 10 C = 90 J
2. a) 9 V
   b) Yes, if the lamps are identical they will be equally bright. If they are different they might draw different current and have different brightness.
1. a) 2 V
   b) No, current is the same everywhere in a series circuit.

6.5 Resistance and resistors

LEARNING SUMMARY	After studying this section, you should be able to: • Describe the effect of resistance. • Use the relationship between resistance, current and voltage. • Explain scientists' model of resistance in metals. • Explain and use Ohm's Law. • Calculate the total resistance of combinations of resistors.

Resistance

AQA	P2	✓
OCR A	P5	✓
OCR B	P4, P6	✓
EDEXCEL	P2	✓
WJEC	P2	✓
CCEA	P2	✓

The components and wires in a circuit **resist** the flow of electric charge. When the **voltage** (or p.d.), V, is fixed, the larger the **resistance** of a circuit the less **current**, I, passes through it.

The resistance of the connecting wires is so small it can usually be ignored. Other metals have a larger resistance, for example the filament of a light bulb has a very large resistance. Metals get hot when charge flows through them. The larger the resistance the hotter they get. A light bulb filament gets so hot that it glows.

For OCR B you need to know that resistance is the gradient of a *V–I* graph.

KEY POINT

Resistance is measured in **ohms** (Ω).

$$\text{Resistance } (\Omega) = \frac{\text{voltage (V)}}{\text{current (A)}}$$

$$R = \frac{V}{I}$$

A model of resistance

AQA	P2	✓
OCR A	P5	✓
OCR B	P6	✓
CCEA	P2	✓

Metals are made of a **lattice** of stationary **positive ions** surrounded by **free electrons**. The moving electrons form the current. In metals with low resistance the electrons require less of a 'push' (p.d) to get through the lattice. The moving electrons **collide** with the stationary ions and make them **vibrate more**. This increase in kinetic energy of the lattice increases the temperature of the metal.

Free electrons   Lattice of positive ions

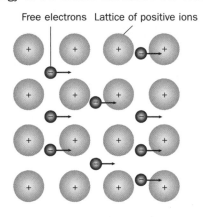

Fixed resistors

AQA	P2	✓
OCR A	P5	✓
OCR B	P6	✓
EDEXCEL	P2	✓
WJEC	P2	✓
CCEA	P2	✓

In some components, such as **resistors** and **metal conductors**, the resistance stays constant when the current and voltage change, providing that the temperature does not change. For this type of fixed resistance if the voltage is increased the current increases so that a graph of current against voltage is a straight line. The current is **directly proportional** to the voltage – doubling the voltage doubles the current. Components that obey this law (**Ohm's Law**) are sometimes called **ohmic** components.

When there is no voltage there is no current, so graphs of *I* against *V* pass through the point (0, 0). Remember this when you are drawing graphs.

A graph of current against voltage for a resistor.

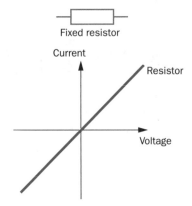

Combining resistors

AQA	P2	✓
OCR A	P5	✓
OCR B	P6	✓
CCEA	P2	✓

Components can be added to a circuit in series or in parallel:

Resistors connected in series and parallel.

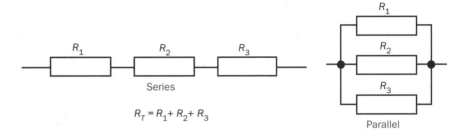

Series

$$R_T = R_1 + R_2 + R_3$$

Parallel

> A series is one after the other, parallel lines are side by side, so series circuits have one component after another and parallel circuits have components that can be drawn side by side instead of one after the other.

For components in series:

- Two (or more) components have more resistance than one on its own.
- The current is the same through each component.
- The p.d. is largest across the component with the largest resistance.
- The p.d.s across the components add up to give the p.d. of the power supply.
- The resistances of all the components add up to give the total resistance of the circuit.

For components in parallel:

- Two (or more) components have less resistance than one on its own.
- The current through each component is the same is if it were the only component.
- The total current will be sum of the currents through all the components.
- The p.d. across all the components will be the same as the power supply p.d.
- The current is largest through the component with the smallest resistance.

> For CCEA and OCRB you need to know that, for resistors in parallel,
> $$\frac{1}{R_T} = \frac{1}{R_1} + \frac{1}{R_2} + \frac{1}{R_3}$$

For components in **series**:

- Two (or more) components in series have more resistance than one on its own. This is because the battery has to push charges through both of them.
- The p.d. is largest across the component with the largest resistance. This is because more work is done by the charge passing through a large resistance than through a small one.

For components in **parallel**:

- A combination of two (or more) components in parallel has less resistance than one component on its own. This is because there is more than one path for charges to flow through.
- The current is largest through the component with the smallest resistance. This is because the same battery voltage makes a larger current flow through a small resistance than through a large one.

6.6 Special resistors

After studying this section, you should be able to:

- Describe how resistance changes with temperature in thermistors and filament lamps.
- Describe how resistance changes with illumination for an LDR
- Describe the resistance of a diode and a variable resistor.
- Explain how these components can be used in circuits.

A filament lamp

AQA	P2	✓
OCR A	P5	✓
OCR B	P6	✓
EDEXCEL	P2	✓
CCEA	P2	✓

The wire in a **filament lamp** gets hotter for larger currents. This increases the resistance so the graph of current against voltage is not a straight line.

A graph of current against voltage for a filament lamp.

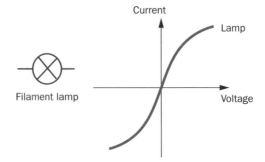

Special resistors

AQA	P2	✓
OCR A	P5	✓
OCR B	P4, P6	✓
EDEXCEL	P2	✓
WJEC	P2	✓
CCEA	P2	✓

A **variable resistor** changes the current in a circuit by changing the resistance. This can be used to change how circuits work. For example to change:

- How long the shutter is open on a digital camera.
- The loudness of the sound from a radio loud speaker.
- The brightness of a bulb.
- The speed of a motor.

Inside one type of variable resistor is a long piece of wire made of metal with a large resistance (called **resistance wire**.) To alter the resistance of the circuit a sliding contact is moved along the wire to change the length of wire in the circuit.

The resistance of a **light dependent resistor (LDR)** decreases as the amount of light falling on it increases. It can be used in a circuit to switch a lamp on, or off, when it gets darker, or lighter.

A graph of resistance against intensity of light for an LDR.

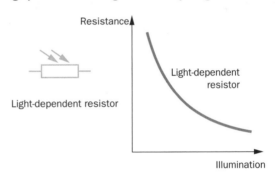

Light-dependent resistor

When an ordinary resistor gets hotter its resistance increases, but for most common thermistors, resistance decreases. When light intensity increases LDRs resistance decreases. The extra energy makes it easier for current to flow in these materials.

The resistance of the most common type of **thermistor** (a negative temperature coefficient (NTC) thermistor) decreases as the temperature increases. It can be used in a circuit to switch a heater or cooling fan, on, or off, at a certain temperature.

A graph of resistance against temperature for a thermistor.

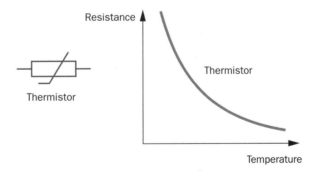

Thermistor

Diodes and LEDs

AQA	P2	✓
OCR B	P6	✓
EDEXCEL	P2	✓

Current will only flow through a **diode** in one direction – the forward direction. In one direction its resistance is very low, but in the other direction, called the **reverse direction**, its resistance is very high.

A **light emitting diode (LED)** is a diode that emits light. LEDs are becoming widely used as low voltage and low energy sources of light.

Notice that symbols for diodes and LEDs (and also LDRs) sometimes include a circle.

> The diode symbols are like arrow heads which show which way the current goes. The little arrows on LEDs and LDRs show whether light is being absorbed or emitted.

LED.

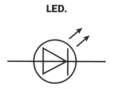

A graph of current against resistance for a diode.

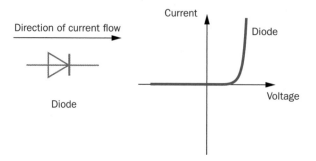

Direction of current flow

Diode

Using thermistors and LDRs

| OCR A | P5 | ✓ |
| OCR B | P6 | ✓ |

Two resistors can be used in a circuit to provide an **output p.d.** with the value that is wanted from a higher **input p.d.** This is called a **potential divider** circuit.

current: $I = \dfrac{V_{in}}{(R_1 + R_2)} = \dfrac{V_1}{R_1} = \dfrac{V_2}{R_2}$

voltage: $V_{in} = V_1 + V_2$

A potential divider circuit.

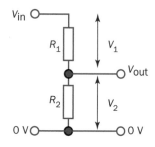

The p.d.s (or voltages) are divided in the same ratio as the resistances.

When a thermistor is used as one of the resistors, the resistance will change with the temperature. This circuit will produce a p.d. that changes with temperature and so it can be used to switch a heater on or off.

An LDR can be used in the same way.

a) The temperature dependent potential divider. b) A light dependent potential divider.

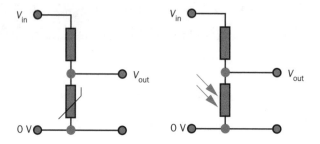

PROGRESS CHECK

1. Why is the resistance of a filament lamp higher when it is switched on?
2. Why would the metal used to make lamp filaments be unsuitable for connecting components in a circuit?
3. What property of a thermistor makes it useful for controlling a heater?
4. Mains electricity is a.c. and computers require a d.c. supply. Which component would be useful in a circuit to connect a computer to use mains electricity? Why?
5. In the potential divider circuit on page 134 if the input p.d. $V_{in} = 5$ V, $R_1 = 200\ \Omega$ and $R_2 = 300\ \Omega$. What is:
   a) V_1
   b) V_2
6. In the circuit with the thermistor on page 134 when the temperature increases what happens to:
   a) The resistance of the thermistor?
   b) The p.d. across the thermistor?
   c) The p.d. across the resistor?

1. Its temperature increases and so resistance increases.
2. It has too high resistance and would get hot when current flowed.
3. Its resistance decreases when temperature increases.
4. A diode – it only allows current to pass in one direction.
5. a) $V_1 = 5\text{ V} \times \frac{200\ \Omega}{(300\ \Omega + 200\ \Omega)} = 2\text{ V}$
   b) $V_2 = 5\text{ V} \times \frac{300\ \Omega}{(300\ \Omega + 200\ \Omega)} = 3\text{ V}$ or $5\text{ V} - 2\text{ V} = 3\text{ V}$
6. a) decreases b) decreases c) increases.

6.7 The mains supply

LEARNING SUMMARY

After studying this section, you should be able to:

- Describe the mains electricity supply in the UK.
- Recall the wiring code for a three-pin plug.
- Explain the role of the fuse and the earth wire.
- Explain the advantages of an RCCB compared to a fuse.
- Use the relationship between power, current, voltage and resistance.

Mains electricity

AQA	P2	✓
OCR A	P5	✓
OCR B	P4	✓
CCEA	P2	✓

KEY POINT

- Mains voltage is 230 V a.c.
- The direction of the current and voltage changes with frequency = 50 Hz
- An electric shock from the mains can kill.

The alternating voltage can be displayed on an **oscilloscope**. The diagram shows an a.c. mains voltage stepped down to 25 V peak voltage

The vertical axis is the voltage of the a.c. supply. One vertical division = 10 V

The horizontal axis is the time. One horizontal division is 10 ms.

The **period** is the time for 1 cycle = 20ms.

$$\text{frequency (Hz)} = \frac{1}{\text{period(s)}} = \frac{1}{20 \text{ ms}} = 50 \text{ Hz}$$

a.c. mains displayed on an oscilloscope

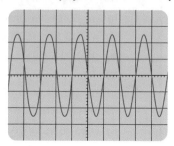

The colour code for mains electricity cables used in buildings and appliances is shown below.

Name of wire	Colour of insulation	Function of the wire
Live	Brown	Carries the high voltage.
Neutral	Blue	The second wire to complete the circuit.
Earth	Green and yellow	A safety wire to stop the appliance becoming live.

This diagram shows the wiring of a three-pin plug for a heater with a metal case. The **fuse** is always connected to the brown, **(live)** wire. A **cable grip** is tightened where the cable enters the plug to stop the wires being pulled out.

A three-pin plug on an earthed appliance.

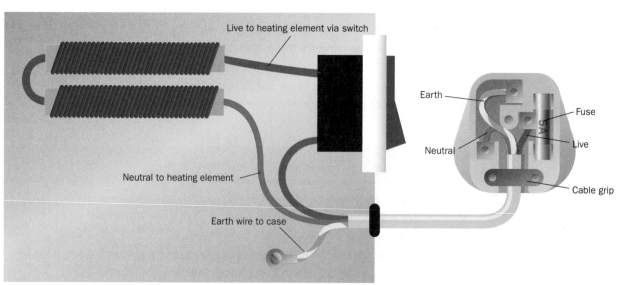

A **fuse** is a piece of wire that is thinner than the other wires in the circuit. If too much current flows it will melt before the wires overheat.

A 3 A fuse will melt if a current of 3 A flows through it. Choose a fuse that is greater than the normal operating current, but as low as possible. If there is a fault, or if too many appliances are plugged into one socket, resulting in a large current, then the fuse will melt and break the circuit preventing a fire.

The **earth wire** is connected to the metal case of appliances so that when they are plugged into the mains supply the metal case is earthed (see page 119.). If there is a fault and the live wire touches the metal case, a very large current flows through the low-resistance path to earth, melting the fuse wire and breaking the circuit.

Double insulated appliances have cases that do not conduct (usually plastic) and have no metal parts that you can touch, so they do not need an earth wire.

Residual current circuit breakers

| AQA | P2 | ✓ |
| OCR B | P4 | ✓ |

The **fuse** takes a short time to melt. It will not prevent you from getting an electric shock if you touch a live appliance. A residual current circuit breaker **(RCCB)** is safer.

Try explaining these safety features to someone. You'll soon find out if you remember them.

These are switches that cut off the electricity very quickly if they detect a difference in the current flowing in the live and the neutral wires, (for example, if the current flows through a person, or appliance casing). Another advantage is that they can be switched back on once the fault is fixed, whereas a fuse must be replaced.

A RCCB can be part of a mains circuit in a building, or a plug-in device that goes between the appliance and the socket. Appliances that are dangerous include:

- those where the cable could get wet, or be cut, for example, lawn mowers and power tools.
- music amplifiers connected to a metal instrument that someone is playing.

Electrical power

AQA	P2	✓
OCR A	P5	✓
OCR B	P4	✓
EDEXCEL	P2	✓
WJEC	P2, P3	✓
CCEA	P2	✓

The power is the rate at which the power supply transfers electrical energy to the appliance. It is measured in watts (W).

$$\text{power (W)} = \frac{\text{energy (J)}}{\text{time (s)}}$$

Electrical power (W) = current (A) × voltage (V)

$P = IV$

Electrical energy (J) = current (A) × voltage (V) × time (s)

$E = IVt$

Example: What is the current in a 2.8 kW kettle?

Using $P = IV$

$I = P \div V$

$I = 2800 \text{ W} \div 230 \text{ V} = 12.2 \text{ A}$

Power and resistance

| OCR B | P6 | ✓ |
| WJEC | P2 | ✓ |

Another useful equation is:

Using $P = IV$ and $R = \dfrac{V}{I}$ so $V = IR$

$P = I \times (IR) = I^2R$

Power (W) = [current(A)]² × resistance (Ω)

$P = I^2R$

Check carefully which equations you are given in the exam and which you need to learn. Make sure you know where to find them on the exam paper. You may find the triangle method useful for rearranging equations.

PROGRESS CHECK

1. Sam replaces a fuse with a piece of high resistance wire. Why is this a bad idea?
2. The earth wire is not connected to a metal appliance. Why is this dangerous?
3. What is the current in:
   a) a 2.5 kW kettle.
   b) a 9 W lamp.
   c) a 300 W TV?
4. Fuses come in 3 A, 5 A and 13 A. Which would you use for each appliance in question 3?
5. Give two advantages of using an RCCB with outdoor Christmas lights.
6. Cables have 100 Ω resistance. Calculate the power wasted heating the cables when the current is:
   a) 0.5 A.
   b) 1 A.

6. a) (0.5 A)² × 100 Ω = 25 W b) (1 A)² × 100 Ω = 100 W
5. If they get wet/wires get cut/other fault the power supply will be cut off. When the fault is fixed the power can be switched back on without replacing the fuse.
4. a) 13 A b) 3 A c) 3 A
3. a) 2500 W ÷ 230 V = 10.9 A b) 9 W ÷ 230 V = 0.04 A c) 300 W ÷ 230 V = 1.3 A
2. If the appliance becomes live it won't melt the fuse, so someone touching it could get a fatal shock.
1. It won't melt if the current gets too high – other wires may melt first causing a fire.

Sample GCSE questions

1 A fan uses mains electricity.

(a) Complete this table that describes the mains electricity supply in the UK. **[3]**

	Mains electricity in the UK
Voltage	230 V
Frequency	50 Hz
a.c. or d.c. ?	a.c.

(b) What is the difference between a.c. and d.c. voltage? A diagram may help you answer. **[2]**

d.c. = direct current, a.c. = alternating current. The difference in the voltages is shown on these graphs.

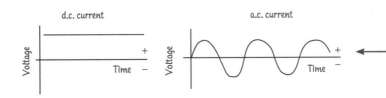

The fan has a power rating of 2 kW. When it is switched on, electric current flows through the circuit.

(c) What is an electric current? **[1]**

It is a flow of electric charge.

(d) Calculate the electric current in the fan. **[2]**

$P = IV$ $P = 2000\,W$ $V = 230\,V$

$I = \dfrac{P}{V}$ $\text{current} = \dfrac{2000\,W}{230\,V} = 8.7\,A$

Current = ...8.7.... A.

(e) The plug has a 13 A fuse. Harry says that a residual current circuit breaker (RCCB) is better. Explain how a fuse works and the advantages of an RCCB. *The quality of your written communication will be assessed in this answer.* **[6]**

A fuse is a thinner piece of wire than the rest of the circuit with a lower melting point. If the current in the circuit is too large the fuse will melt before the other wires and break the circuit. This will prevent overheating and fires, but it is too slow to prevent electric shocks. The advantages of the RCCB are that it will stop the current much faster than a fuse and it can be reset after the fault is fixed, whereas the fuse has to be replaced.

[Total = 14]

Graphs must have axes labelled 'voltage' and 'time'. Alternatively, write 'd.c. is direct current so the voltage is always in the same direction. a.c. is alternating current so the voltage keeps changing direction.' (Note: not just changing, but changing direction.)

Write down the equation and the values you are using. You may get marks. Use the value for voltage that you used in part (a) if it is incorrect you will not lose marks again – your answer here will be marked correct.

Marks will be awarded depending on the number of relevant points included in the answer and the spelling, punctuation and grammar. In this question there are 8 or 9 relevant points so 7 or 8 including 2 advantages of RCCBs with good spelling punctuation and grammar will gain full marks.

Sample GCSE questions

2 Kate is investigating how the thickness of a piece of wire affects its resistance.

(a) Complete this circuit diagram to show how she should connect an ammeter and a voltmeter to measure the voltage across the wire and the current flowing through it. **[2]**

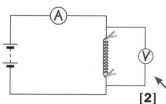

You must use the correct circuit symbols, leave no gaps, and connect the voltmeter across the wire and the ammeter in line with it.

(b) Kate has a selection of wires of different thickness. State two factors that must be the same for all the test wires. **[2]**

　　1.　　The length of the wire.

　　2.　　The metal the wire is made from.

(c) Here are Kate's results.

Thickness of wire (mm)	Voltmeter reading (volts)	Ammeter reading (amps)	Resistance (ohms)
0.5	12.0	0.30	40
1.0	12.0	0.67	18
1.5	12.0	1.20	10
2.5	12.0	2.67	4.5
3.5	12.0	4.80	

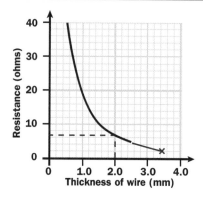

Remember to include the unit of resistance, Ω or 'ohms' is OK for the mark.

(i) Calculate the resistance of the 3.5 mm thickness wire. **[2]**

$R = V/I$　$R = 12.0V \div 4.80A = 2.5\ \Omega$

Resistance = ...2.5Ω....

The graph is a smooth curve so do not draw a straight line with a ruler. Use a pencil to continue the curve. Turn the paper if it helps. Do not make the line very thick or draw lots of attempts. Remember that exam papers are scanned and marked on screen, so a poorly erased line may still appear.

(ii) Plot your calculated value on the graph. **[1]**

(iii) Extend the line to include your plotted value. **[1]**

(d) Use the graph to find the resistance of a wire with thickness 2.0 mm. **[1]**

Resistance = ...6.5Ω....

[Total = 9]

Exam practice questions

1 Gemma uses an electrostatic paint spray to paint a metal fence. The spray nozzle gives the paint droplets a positive charge.

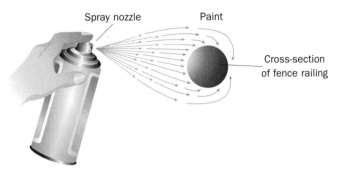

Spray nozzle   Paint

Cross-section of fence railing

(a) The metal fence is given an electric charge. Explain whether this is a positive or negative charge. **[2]**

....

....

(b) Explain how charging the paint and the fence improves the paint spraying process. *The quality of your written communication will be assessed in this answer.* **[6]**

....

....

....

....

....

[Total = 8]

2 Explain how a defibrillator is used to restart the heart. *The quality of your written communication will be assessed in this answer.* **[6]**

....

....

....

....

3 A car headlamp uses a 12 V battery.

(a) 300 C of charge pass through the bulb in 1 minute. Calculate the current in the lamp. **[2]**

....

....

Exam practice questions

(b) Calculate the energy transferred to the lamp by 300 C of charge flowing through it. **[2]**

....

....

(c) Calculate the power of the lamp. **[2]**

....

....

[Total = 6]

4 Sadia has bought a new electric lamp. It is double insulated so it does not need earthing.

(a) What does double insulated mean? **[1]**

....

(b) The lamp has a 3 A fuse. Describe what happens when **[2]**

(i) the normal current of 0.25 A flows in the circuit.

....

(ii) there is a fault and the current increases to 5 A.

....

(c) The fault is fixed and the fuse is replaced with a 13 A fuse. Explain why this is not a good idea. **[2]**

....

....

[Total = 5]

5 A mains electricity fan heater has two switches (mains is 230 V).

Switch X turns the fan motor on and off. Switch Y turns the heater on and off.

(a) Label the switches X and Y on the circuit diagram. **[1]**

(b) Is it possible to turn the heater on without the fan? .... **[1]**

(c) Is it possible to turn the fan on without the heater? .... **[1]**

(d) Explain why the switches have been arranged to work in this way. **[3]**

....

....

....

Exam practice questions

(e) The resistance of the motor is 70 Ω. Calculate the current in the motor. **[3]**

....

....

(f) The current in the heater is 0.6 A. What is the total current from the power supply? **[3]**

....

....

[Total = 12]

6 This diagram shows a circuit that can be used to control a lighting circuit.

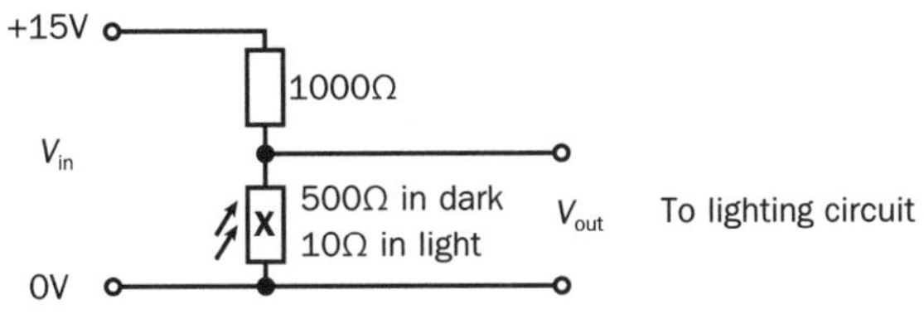

+15V —————————
                | 1000Ω
V_{in} |———————o
        /|X| 500Ω in dark V_{out} To lighting circuit
        /|X| 10Ω in light
0V ——————————————o

(a) What is the name of the component marked X in the circuit? **[1]**

....

(b) In the dark, what is the total resistance of the circuit? **[1]**

....

(c) In the dark, what is the potential difference: **[2]**

(i) across the 1000 Ω resistor? ....

(ii) across the component marked X? ....

(d) Describe what happens to the resistance and the potential difference across component X as the light level changes. *The quality of your written communication will be assessed in this answer.* **[6]**

....

....

....

(e) Explain how this could be used to control the lighting circuit. **[2]**

....

....

[Total = 9]

7 Radioactivity

The following topics are covered in this chapter:

- **Atomic structure**
- **Radioactive decay**
- **Living with radioactivity**
- **Uses of radioactive material**
- **Nuclear fission and fusion**

7.1 Atomic structure

LEARNING SUMMARY

After studying this section, you should be able to:

- Describe the atom and the particles it is made of.
- Write symbols for nuclei.
- Complete nuclear equations for alpha and beta decay.
- Describe, and explain the results of, Rutherford's gold foil experiment.

The atom

AQA	P2	✓
OCR A	P6, P7	✓
OCR B	P4	✓
EDEXCEL	P2, P3	✓
WJEC	P2	✓
CCEA	P1	✓

Atoms are about 10^{-10} m or 0.1 nanometres in diameter. Neutral atoms have the same number of protons and electrons. When ionisation occurs atoms gain or lose electrons, becoming negatively or positively charged **ions**.

> **KEY POINT**
>
> - The **atom** is mostly empty space with almost all the mass concentrated in the small positively charged nucleus at the centre.
> - The nucleus is very small compared to the volume, or shell, around the nucleus that contains the **electrons**.

A model of the atom.

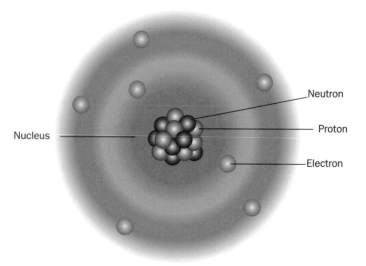

Nucleus

Neutron

Proton

Electron

Particles in the atom	Symbol	Where found in the atom	Relative mass	Relative charge
Proton	p	In nucleus	1	+1
Neutron	n	In nucleus	1	0 (neutral)
Electron	e	Outside nucleus	$\frac{1}{1840}$	−1

Make a set of flash cards with these words on one side and what they mean on the other. Keep looking at them and this will help you remember them.

> **KEY POINT**
>
> The **nucleus** of an atom contains particles called **protons** and **neutrons**. These are also called **nucleons**.
>
> The **atomic number** or **proton number**, **Z**, is the number of protons in the nucleus.

The number of protons is what makes the atom into the element it is, so, for example, hydrogen always has one proton and carbon always has six.

> **KEY POINT**
>
> The **mass number** or **nucleon number**, **A**, is the total number of protons and neutrons in the nucleus.
>
> **Isotopes** of an element have the same number of protons in the nucleus, but different numbers of neutrons. This means that different isotopes of an element each have the same proton or atomic number but a different nucleon or mass number.

Isotopes of the same element have exactly the same chemical properties, but they have different mass and nuclear stability. For example, carbon-12 is a stable isotope of carbon with 6 protons and 6 neutrons and carbon-14 is a radioactive isotope with 6 protons and 8 neutrons.

Nuclear equations

AQA	P2	✓
OCR A	P6, P7	✓
OCR B	P4	✓
EDEXCEL	P2, P3	✓
WJEC	P2	✓
CCEA	P1	✓

Nuclei are given symbols, for example, this is the symbol for the stable isotope carbon-12 which has 6 protons and 6 neutrons:

$$^{12}_{6}\text{C}$$

This is the symbol for an alpha particle (see page 147), which is the same as a helium nucleus. Sometimes α is used instead of He:

$$^{4}_{2}\text{He}$$

A beta particle (see page 147) is not a nucleus, but this symbol is used for a beta particle in a nuclear equation. Sometimes β is used instead of e:

$$^{0}_{-1}\text{e}$$

> **KEY POINT**
>
> Before and after a nuclear decay or reaction:
>
> - The total of the mass numbers must be the same.
> - The total of the atomic numbers must be the same.

Example: Alpha decay of radon-220.

$$^{220}_{86}\text{Rn} \rightarrow \, ^{216}_{84}\text{Po} + \, ^{4}_{2}\text{He}$$

$220 = 216 + 4$ and $86 = 84 + 2$

Example: Beta decay of carbon-14.

$$^{14}_{6}\text{C} \rightarrow \, ^{14}_{7}\text{N} + \, ^{0}_{-1}\text{e}$$

$14 = 14 + 0$ and $6 = 7 + (-1)$

A model of the atom

| OCR A | P6 | ✓ |
| CCEA | P1 | ✓ |

Before 1910 scientists had a **plum pudding model** of the **atom**. In this model the atom is described as being made of positively charged material (the pudding) with negatively charged electrons (the plums) inside.

> This is an example of how scientists change their ideas over time.

Later Ernest Rutherford investigated the structure of atoms by firing **alpha particles** at **gold foil**. He suggested the **nuclear model** of the atom.

What happened to the alpha particles	Ernest Rutherford's explanation – the nuclear atom
The majority went straight through the foil, without being deflected.	The atom is mostly empty space.
Some were deflected and there was a range deflection angles.	Parts of the atom have positive charge.
A very small number were 'back-scattered' – they came straight back towards the alpha particle source.	There is a tiny region of concentrated mass and positive charge which repels the very small number of alpha particles that have a head-on collision.

The experiment was carried out in a vacuum, so that the alpha particles were not stopped by the air. A fluorescent screen flashed each time an alpha particle hit it. The location of each flash showed whether the alpha particle had travelled straight through the foil or whether it had been deflected. Hans Geiger and Ernest Marsden counted small flashes of light under the direction of Ernest Rutherford. The experiment is often referred to as Rutherford scattering.

> Remember: Most straight through, some deflected, very few straight back.

Alpha particle scattering by gold foil.

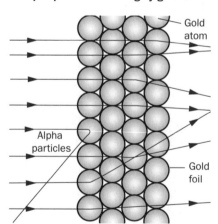

Gold atom

Alpha particles

Gold foil

PROGRESS CHECK

1. Which particles are found in the:
   a) atom
   b) nucleus?
2. The atomic number of oxygen is 8. What does this tell you?
3. What evidence did Rutherford find for the atom being mostly empty space?
4. Write an equation for:
   a) The alpha decay of uranium-235 (symbol= U, Z = 92) to thorium (symbol = Th).
   b) The beta decay of nitrogen-16 (symbol= N, Z = 7) to oxygen (symbol = O).

b) $^{16}_{7}N \rightarrow ^{16}_{8}O + ^{0}_{-1}e$

4. a) $^{235}_{92}U \rightarrow ^{231}_{90}Th + ^{4}_{2}He$

3. Most of the alpha particles went straight through.

2. There are 8 protons in the nucleus (and 8 orbital electrons when the atom is neutral)

1. a) protons, neutrons, electrons b) protons, neutrons.

7.2 Radioactive decay

LEARNING SUMMARY

After studying this section, you should be able to:

- Describe the properties of alpha, beta and gamma emissions.
- Explain that radioactive decay is random.
- Explain and calculate half-life.
- Understand the term 'activity' in relation to a radioisotope.
- Interpret decay curves of isotopes.

Alpha, beta and gamma

AQA	P2	✓
OCR A	P6	✓
OCR B	P2, P4	✓
EDEXCEL	P2, P3	✓
WJEC	P1, P2	✓
CCEA	P1	✓

There are three main types of radioactive emissions: **alpha particles**, **beta particles** and **gamma rays**. This table shows some of the properties of the different types of radiation.

Radiation	Ionising effect	Electric charge	Stopped by ...	Affected by electric and magnetic fields?
Alpha (α)	Strong	+	Skin. A few cm of air. A sheet of paper.	Yes
Beta (β)	Weak	−	A thin sheet of aluminium or other metal.	Yes
Gamma (γ)	Very weak	Neutral	A thick lead sheet reduces intensity. Thick concrete blocks reduce intensity.	No

When alpha and beta particles are emitted the nucleus changes into a different element. When gamma rays are emitted the element does not change.

KEY POINT

Alpha emission is when two protons and two neutrons leave the nucleus as one particle, called an **alpha particle**. An alpha particle is identical to a helium nucleus.

Beta emission is when a neutron decays to a proton and an electron inside the nucleus. The high energy electron leaves the nucleus as a **beta particle**.

Gamma emission is when the nucleus emits a short burst of high-energy **electromagnetic radiation**. The **gamma ray** has a high frequency and a short wavelength.

| Alpha particle (α) | Beta particle (β) | Gamma rays (γ) |

Radioactive decay

AQA	P2	✓
OCR A	P6	✓
OCR B	P2, P4	✓
EDEXCEL	P2	✓
WJEC	P1, P2	✓
CCEA	P1	✓

KEY POINT

A radioactive material contains nuclei that are unstable and emit nuclear radiation. This process is called **radioactive decay**. Radioactive decay is **random**.

It is not possible to predict when it will happen, nor is it possible to make it happen by a chemical or physical process, for example by heating the material. A radioactive source contains millions of atoms. The number of radioactive emissions a second depends on two things:

- The type of nucleus – some combinations of protons and neutrons are more stable than others.
- The number of undecayed nuclei in the sample – with double the number of nuclei, on average, there will be double the number of emissions per second.

Over a period of time the **activity** of a source gradually reduces.

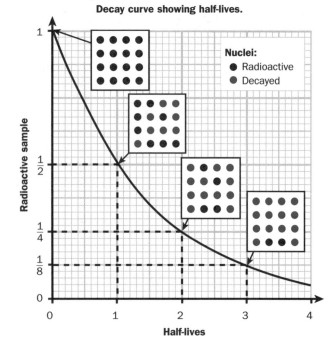

Decay curve showing half-lives.

Nuclei:
● Radioactive
● Decayed

The **half-life** of an isotope is the average time taken for half of the nuclei present to decay.

Example: Technetium-99m (Tc-99m) decays by gamma emission to Technetium-99 (Tc-99) with a half-life of six hours. After six hours, only half of the Tc-99m nuclei remain. After another six hours, that is a total of 12 hours, only one quarter are left.

This pattern is the same for all isotopes, but the value of the half-life is different. Carbon-14 has a half-life of 5730 years; some isotopes have a half-life of less than a second.

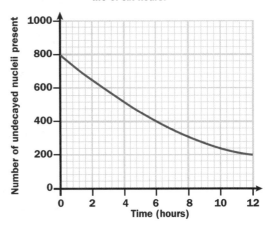

The decay of a sample radioactive nuclei with a half-life of six hours.

Activity

The number of radioactive emissions a second is called the **activity** of the source. On average, in one half-life the activity of a source will reduce to one half of its original value.

Common mistakes are to say:
- After three half-lives there are 1/3 or 1/6 of the radioactive nuclei left – but it is 1/8.
- '1/8 of the nuclei have decayed' – but it is 7/8 because only 1/8 are left.

After this number of half-lives ...	The activity has dropped to this fraction of the initial value ...	Which is the same as ...
1	$\frac{1}{2}$	$\frac{1}{2^1}$
2	$\frac{1}{2} \times \frac{1}{2} = \frac{1}{4}$	$\frac{1}{2^2}$
3	$\frac{1}{2} \times \frac{1}{2} \times \frac{1}{2} = \frac{1}{8}$	$\frac{1}{2^3}$
10	$\frac{1}{1024}$	$\frac{1}{2^{10}}$

For a source with a half-life of 2 hours, after 20 hours the activity is ...

20 hours ÷ 2 hours = 10 half-lives

Fraction of original activity = $\frac{1}{2^{10}}$ = 1/1024

After ten half-lives, the activity has dropped to less than one thousandth of the original activity. This is often used as a measure of the time for a sample to decay to a negligible amount.

PROGRESS CHECK

1. Answer alpha, beta or gamma to the following.
   a) Which type of radiation has a positive charge?
   b) Which types of radiation can be stopped by a thin sheet of aluminium?
   c) Which type of radiation passes through a sheet of paper and is deflected by a magnet?

PROGRESS CHECK

2️⃣ A radioactive source has a half-life of 24 hours. What fraction will remain after **a)** one day and **b)** four days?

3️⃣ Describe what happens to an alpha particle as it passes through acid.

4️⃣ A radioactive source with a half-life of three hours has an activity of 16 000 Bq. What is the activity after:

a) 3 hours?

b) 9 hours?

c) 30 hours?

d) When will the activity be 500 Bq?

1. a) alpha particle
   b) alpha particles and beta particles
   c) beta
2. a) ½ b) 1/16
3. It loses energy as it slows down and attracts 2 electrons, becoming a helium atom.
4. a) 16 000 ÷ 2 = 8000 Bq b) 16 000 ÷ 2^3 = 2000 Bq c) 16 000 ÷ 2^{10} = 16
   d) 16 000 Bq × $\frac{1}{2}$ × $\frac{1}{2}$ × $\frac{1}{2}$ × $\frac{1}{2}$ × $\frac{1}{2}$ = 500 Bq = 5 half lives = 5 × 3 hours = 15 hours

7.3 Living with radioactivity

LEARNING SUMMARY

After studying this section, you should be able to:

- Understand and explain the term background radiation.
- Describe some sources of background radiation.
- Explain the difference between contamination and irradiation.
- Understand the term radiation dose.
- Describe the different risks of α, β and γ exposure.

Background radiation

AQA	P2	✓
OCR A	P6	✓
OCR B	P4	✓
EDEXCEL	P2	✓
WJEC	P1	✓
CCEA	P1	✓

Some radioactive materials occur naturally, others are man-made. **Cosmic rays** from space make some of the carbon dioxide in the atmosphere radioactive. The carbon dioxide is used by plants and enters food chains. This makes all living things radioactive. Some rocks are naturally radioactive.

KEY POINT

We receive a low level of radiation from these sources all the time. It is called **background radiation**.

Background radiation comes from:

- Radon (a radioactive gas) from rocks.
- Soil and building materials.
- Medical and industrial uses of radioactive materials.
- Food and drink.
- Cosmic rays (from outer space).
- 'Leaks' from radioactive waste and nuclear power stations.

Background radiation in the UK.

Cosmic rays 10%

Other 0.2%

Food and drink 11.5%

Nuclear power and weapons 0.3%

Medical 14%

Ground and buildings 14%

Radon gas 50%

Sources of background radiation: cosmic rays are from outer space (not the Sun).

Food and drink are radioactive as explained on page 150 – not because of food irradiation.

Some rocks are more radioactive than others, so the level of background radiation can depend on the underlying rocks. In certain areas, **radon** from some rocks can build up in houses. It emits alpha radiation, so it is particularly damaging in the lungs. Houses with high levels of radon can have under-floor fans fitted to keep the radon out of the house.

In some parts of the country the rocks are more radioactive than in others and there is a higher level of background radiation.

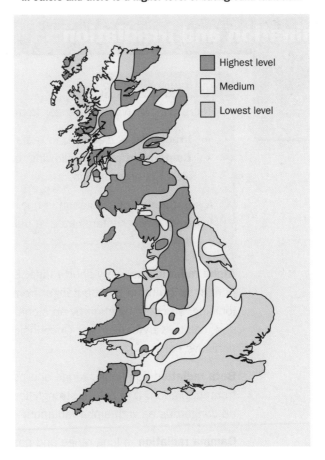

Highest level

Medium

Lowest level

Background radiation is the name given to radioactive emissions from nuclei in our surroundings. Do not confuse this with radiation from mobile phones or cosmic microwave background radiation.

Dangers of radiation

AQA	P2	✓
OCR A	P6	✓
OCR B	P2, P4	✓
EDEXCEL	P2, P3	✓
WJEC	P1	✓
CCEA	P1	✓

KEY POINT

Ionisation is when atoms gain or lose electrons becoming negatively or positively charged. **Ionising radiation** is radiation that has enough energy to ionise the atoms in molecules. The **ions** can take part in chemical reactions. In **living cells** this can damage or kill them. It can damage the **DNA** so that the cell **mutates** into a **cancer** cell.

It is not possible to predict which cells will be damaged by exposure to radiation or who will get cancer.

Scientists studied the survivors of incidents where people were exposed to ionising radiation. They measured the amount of exposure and recorded how many people later suffered from cancer. The risk of cancer increases with increased exposure to radiation. People tend to overestimate the risk from radiation because it is invisible and unfamiliar. They underestimate the risk of familiar activities, like smoking.

Contamination and irradiation

AQA	P2	✓
OCR A	P6	✓
OCR B	P4	✓
EDEXCEL	P2, P3	✓
WJEC	P1	✓
CCEA	P1	✓

KEY POINT

There are two types of danger from radioactive materials.

- **Irradiation** is exposure to radiation from a source outside the body.
- **Contamination** is swallowing, breathing in, or getting radioactive material on your skin.

A short period of irradiation is not as dangerous as being contaminated because, once contaminated, a person is continually being irradiated.

Alpha radiation has very short range. Even if it reaches the skin it does not penetrate, so there is little danger from irradiation. However, it is strongly ionising, so contamination by an alpha source, for example breathing in radon gas, can be very dangerous. Once inside the lungs it will keep irradiating sensitive cells.

Beta radiation has longer range and penetrates the skin so there is more danger from irradiation. It is not as strongly ionising as alpha, so contamination is not as dangerous as with alpha radiation.

Gamma radiation is long range and passes right through the body. There is a danger from irradiation if the radiation levels are high. It is only weakly ionising and most of the rays pass right through the body without hitting anything so contamination is less dangerous than with alpha or beta radiation.

Radiation dose

OCR A	P6	✓
EDEXCEL	P3	✓
WJEC	P1	✓

Radiation dose, measured in **sieverts** (Sv), is a measure of the possible harm done to the body. Radiation dose depends on the type of radiation, the time of exposure, and how sensitive the tissue exposed is to radiation. The dose is linked to the risk of cancer developing. Alpha radiation is strongly ionising and so the dose is twenty times larger from alpha than from beta or gamma radiation.

The normal UK background dose is a few milliSieverts. A fatal dose is between 4 Sv and 5 Sv given in one go.

> **PROGRESS CHECK**
>
> 1. Give a source of background radiation.
> 2. How much of the background radiation in the UK is from radon gas?
> 3. Give one effect of ionising radiation on living cells.
> 4. 'She has been irradiated, she will get cancer.' What is wrong with this statement?
> 5. Name a part of the UK that has high background radiation.
>
> 1. One of sources on piechart on page 151, e.g. cosmic rays (outer space)
> 2. 50% (or half)
> 3. kill, damage, damage DNA, turn cancerous
> 4. Her risk of getting cancer is increased, but we can't tell whether she will get cancer.
> 5. e.g. Cornwall, Cumbria, other orange areas on map

7.4 Uses of radioactive material

LEARNING SUMMARY	**After studying this section, you should be able to:**
	• Explain whether alpha, beta or gamma radiation is the most suitable to use.
	• Choose the most suitable radioisotope for a job.
	• Describe how radioisotopes are used.
	• Consider benefits and risks in the use of radioactive isotopes.

Choosing the best isotope

AQA	P2	✓
OCR A	P6	✓
OCR B	P4	✓
EDEXCEL	P2, P3	✓
WJEC	P2	✓
CCEA	P1	✓

Radioactive materials can be useful as well as harmful. When a radioactive material is to be used certain factors need to be considered.

- Alpha, beta or gamma radiation is chosen depending on the **range** and the **absorption** required.
- An isotope is chosen depending on:
  - whether it emits alpha, beta or gamma radiation
  - how long it remains radioactive, which depends on the half-life.

Radioactive isotopes for different uses are produced in nuclear reactors.

Medical tracers

AQA	P2	✓
OCR A	P6	✓
OCR B	P2, P4	✓
EDEXCEL	P2, P3	✓
WJEC	P2	✓
CCEA	P1	✓

Isotopes which emit **gamma radiation** (or sometimes **beta radiation**) are used in **medical tracers**. The patient drinks, inhales, or is injected with the tracer, which is chosen to collect in the organ doctors want to examine.

Example: Radioactive iodine is taken up by the thyroid gland, which can then be viewed using a **gamma camera** that detects the gamma radiation passing out of the body. The tracer must not decay before it has moved to the organ being investigated, but it must not last so long that the patient stays radioactive for weeks afterwards.

Treating cancer

AQA	P2	✓
OCR A	P6	✓
OCR B	P2, P4	✓
EDEXCEL	P2, P3	✓
WJEC	P2	✓
CCEA	P1	✓

Isotopes which emit a higher dose of **gamma radiation** than tracers are used to build up in the cancer and kill the cancer cells. Alternatively, beams of **gamma rays** are concentrated on a tumour to kill the cancer cells.

Example: Cobalt-60 emits high energy gamma rays and remains radioactive for years.

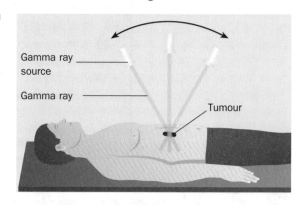

Treating cancer.

Gamma ray source

Gamma ray

Tumour

Benefits and risks

AQA	P2	✓
OCR A	P6	✓
OCR B	P2, P4	✓
EDEXCEL	P2, P3	✓
WJEC	P2	✓
CCEA	P1	✓

When radioactive materials are used, we have to decide whether the **benefits** outweigh the **risks**.

Example: Treatment benefits patients, but does not benefit hospital staff who work with radioactive materials regularly.

Radiation workers have their exposure monitored and take safety precautions to keep their dose as low as possible:

- They wear protective clothing.
- They keep a long distance away, for example they use tongs to handle sources.
- They keep the exposure time short.
- They shield sources and label them with

> Use a mnemonic that contains the first letters of each to remember safety precautions: <u>M</u>y <u>c</u>urtains <u>d</u>im the <u>b</u>right <u>l</u>ight
>
> <u>M</u>onitor, <u>c</u>lothing, <u>d</u>istance, <u>t</u>ime, <u>b</u>arrier, <u>l</u>abel.

Radioactive hazard symbol.

Sterilisation

AQA	P2	✓
OCR A	P6	✓
OCR B	P2, P4	✓
EDEXCEL	P2	✓
WJEC	P2	✓
CCEA	P1	✓

Gamma radiation destroys microbes and is used for:

- Sterilising equipment, for example surgical instruments.
- Food irradiation to extend the shelf-life of perishable food.

The food or equipment does not become radioactive because it is only **irradiated**, there is no **contamination**. It is irradiated inside the plastic packaging, so that it stays sterile.

Smoke detectors

OCR B	P2, P4	✓
EDEXCEL	P2	✓
CCEA	P1	✓

Isotopes which emit **alpha radiation** are used in **smoke detectors**. The alpha radiation crosses a small gap and is picked up by a detector. If smoke is present, the alpha radiation is stopped by the smoke particles. No radiation reaches the detector and the alarm sounds.

Beta and gamma radiation are unsuitable because they pass through the smoke.

Tracers

OCR B	P2, P4	✓
EDEXCEL	P2	✓
CCEA	P1	✓

Isotopes which emit beta radiation or gamma radiation can be used as tracers. Because a tracer is radioactive, detectors can track where it goes. A tracer can be added to sewage at an ocean outlet, or as it enters a river, to trace its movement. In this way, leaks in power station heat exchangers can be tracked. The isotope used has an activity that will fall to zero quickly after the test is done.

Thickness detectors (gauging)

OCR B	P2	✓
EDEXCEL	P2	✓

Isotopes which emit beta radiation are used in thickness detectors.

Example: Some of the beta radiation is absorbed by a paper sheet. If the sheet is too thick, less beta radiation is detected and the pressure of the rollers is increased. If the sheet is too thin, more beta radiation is detected and the pressure is reduced.

Using a beta source to control paper thickness.

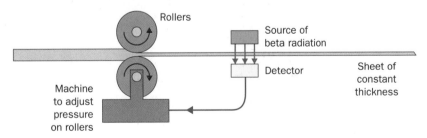

Carbon dating

| OCR B | P4 | ✓ |
| WJEC | P2 | ✓ |

The amount of radioactive carbon left in old materials that were once living can be used to calculate their age.

Carbon dating:

- Can only be used for things that once lived.
- Cannot be used for objects older than 10 half-lives = 10 × 5730 years or less than 100 years.

Carbon dating.

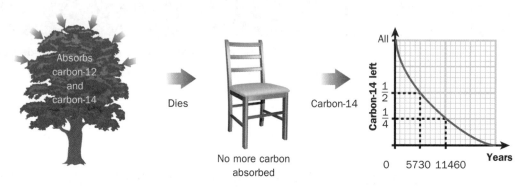

Some rocks contain a radioactive isotope of uranium that decays to lead, so they can be dated by comparing the amounts of uranium and lead. The more lead there is the older the rock is.

Learn the properties of alpha, beta and gamma radiation, so that you can explain why each is chosen.

PROGRESS CHECK

1. Give a use of:
   a) Alpha radiation.
   b) Beta radiation.
   c) Gamma radiation.
2. Explain why healthy tissue is not killed by the gamma rays from the cobalt-60 during cancer treatment.
3. Why are alpha and gamma radiation not suitable for gauging the thickness of paper?
4. Why is carbon dating not used to confirm the age of a 60 000 year-old egg?

1. a) e.g. smoke detector
   b) e.g. medical tracer
   c) e.g. cancer treatment
2. Each of the beams in the diagram is a low dose so it doesn't kill the cells, but when combined in the tumour the dose is high enough to kill the cells.
3. Alpha is all absorbed by paper, gamma will all pass through the paper.
4. There is no radioactive carbon left in the egg, so it could be much older – you can't tell.

7.5 Nuclear fission and fusion

LEARNING SUMMARY

After studying this section, you should be able to:

- Describe nuclear fission.
- Recall $E = mc^2$.
- Describe how chain reactions are used in nuclear reactors.
- Describe how nuclear waste is disposed of.
- Describe nuclear fusion.

Nuclear fission

AQA	P2	✓
OCR A	P6	✓
OCR B	P2, P4	✓
EDEXCEL	P2	✓
WJEC	P2, P3	✓
CCEA	P1	✓

KEY POINT

Nuclear fission is when a nucleus splits into two nuclei of about equal size, and two or three neutrons.

Example: After uranium-235 absorbs a neutron:

A neutron absorbed by a uranium nucleus causes a nuclear fission.

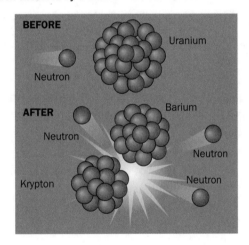

$E = mc^2$

OCR A	P6, P7	✓
EDEXCEL	P3	✓
WJEC	P3	✓

KEY POINT

A small amount of mass is converted into a large amount of energy. The energy is calculated using Einstein's equation:

Energy (J) = mass (kg) × speed of light in a vacuum (m/s)^2

$E = mc^2$

About a **million times more energy** is released than in a chemical reaction.

A chain reaction

AQA	P2	✓
OCR A	P6	✓
OCR B	P4	✓
EDEXCEL	P2	✓
WJEC	P2	✓
CCEA	P1	✓

KEY POINT

After the new neutrons slow down, they can strike more uranium nuclei and cause more fission events. This is called a **chain reaction**.

A chain reaction.

uranium-239 nucleus

neutron

If the chain reaction runs out of control it is an **atomic bomb**. In a **nuclear reactor** the process is controlled. In a nuclear power station the energy released is used to generate electricity.

Nuclear reactors

AQA	P2	✓
OCR A	P6	✓
OCR B	P4	✓
EDEXCEL	P2	✓
WJEC	P2	✓
CCEA	P1	✓

In a nuclear reactor:

- The **fuel rods** are made of **uranium-235** or **plutonium-239**.
- The **moderator** is a material that slows down the neutrons.
- The energy heats up the reactor core.
- A **coolant** is circulated to remove the heat.
- The hot coolant is used to heat water to steam, to turn the power station turbines.
- The **control rods** are moved into the reactor to absorb neutrons to slow or stop the reaction.

Nuclear reactor.

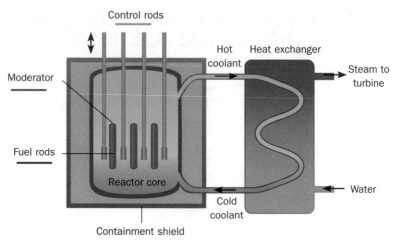

Waste disposal

OCR A   P6   ✓
OCR B   P2, P4 ✓
EDEXCEL   P2   ✓
CCEA   P1   ✓

KEY POINT

Reactors produce **radioactive waste**. The half-life of some isotopes is thousands of years, so radioactive waste must be kept safely contained for thousands of years.

You may be asked for advantages and disadvantages, for example, of methods of waste disposal or nuclear reactors. To gain full marks you must give at least one advantage *and* one disadvantage.

If you are asked for your opinion it doesn't matter if you say 'yes' or 'no', but you must say one or the other. Remember that most of the marks are for justifying your choice.

Types of radioactive waste	Examples	Disposal
Low level waste	Used protective clothing.	Sealed into containers. Put into landfill sites.
Intermediate level waste	Material from reactors.	Mixed with concrete. Stored in stainless-steel containers.
High level waste	Used fuel rods.	Kept under water in cooling tanks (it decays so fast it gets hot). Eventually becomes intermediate level waste.

Where is the safest place to store the radioactive waste?

- At the bottom of the sea, but the containers may leak.
- Underground, but the containers may leak and earthquakes or other changes to the rocks may occur.
- On the surface, but needs guarding from terrorists, for thousands of years.
- Blast into space, but there is a danger of rocket explosion.

Nuclear fusion

AQA   P2   ✓
OCR A   P6, P7   ✓
OCR B   P4   ✓
EDEXCEL   P2, P3   ✓
WJEC   P2, P3   ✓
CCEA   P1   ✓

KEY POINT

When two small nuclei are close enough together they can **fuse** together to form a larger nucleus. This releases a large amount of energy.

Draw up a table to compare fusion and fission. It will help you when you revise.

The problem is getting the two nuclei close enough, because nuclei are positively charged and **repel** each other. Inside stars the temperatures are high enough for the nuclei to have enough energy to get close enough for nuclear fusion to occur.

Scientific research is continuing to try and control the nuclear fusion reaction and produce nuclear fusion reactors. Advantages of a fusion reactor over a fission reactor would be:

- The reaction stops if there is a fault.
- There would be less radioactive waste.
- More energy would be produced.

KEY POINT

The protons and neutrons inside the nucleus are held together by a force called the **strong force**.

The problem is getting, and keeping, the temperature and pressure high enough to overcome the repulsive force, so that the nuclei get close enough for the strong force to take over. When fusion happens a small amount of mass is converted into a large amount of energy ($E = mc^2$).

The ITER project is to build a fusion reactor in which hydrogen-1 and hydrogen-2 nuclei are fused to give helium-3 nuclei. It involves many scientists from 20 countries. Scientists work together because the project is so expensive and because they make more progress if they share ideas.

Cold fusion

AQA	P2	✓
OCR A	P6, P7	✓
OCR B	P4	✓
EDEXCEL	P2, P3	✓
WJEC	P2, P3	✓
CCEA	P1	✓

Fusion research and reactors are very expensive. Martin Fleischmann and Stanley Pons were scientists with a track record of success in other fields. In 1989 they announced that they had achieved nuclear fusion without high temperatures in a process called **cold fusion**.

They used a cheaper electrolysis experiment with a palladium and a platinum electrode in 'heavy water.' This is water made from the hydrogen isotope that has a neutron in the nucleus. A lot of heat was generated and they claimed to have detected neutrons showing that fusion had taken place.

> This is an example of 'How Science Works' which is on all specifications, although the details of the cold fusion experiment are only required for OCR B and Edexcel.

Many scientists tried to **replicate** the experiment. Fleischmann and Pons released very few details, possibly because of the chance of a Nobel prize, or a patent being awarded. (If they released too many details someone else might get there first.) They had rushed to publish something because they thought one of the scientists who **peer reviewed** their request for research money might claim credit for their idea. Some scientists claimed to have replicated their experiment, and some could not. Eventually almost all the claims of success were retracted as careful checking showed the heat and neutrons were not due to nuclear fusion. Fleischmann and Pons could have brought more details of the experiment, or data as evidence, but they never did. Scientists assume they could not **reproduce** it.

> Even a scientist who usually shares their ideas may be afraid that others will not treat him, or her, the same way when there is money or a Nobel Prize at stake.

The **peer review** process works best when scientists behave ethically and for the common good. Cold fusion is an example where the peer review process did not work well, because scientists did not share their results and ideas.

PROGRESS CHECK

1. What is the difference between nuclear fission and nuclear fusion?
2. Describe a chain reaction in plutonium-239.
3. Explain what control rods are used for.
4. How are fuel rods disposed of when the fuel is used up?
5. How much energy results from 0.1 g of fuel being converted to energy?
6. What is the force that holds protons and neutrons together?

1. Fission is splitting large nuclei into two roughly equal parts, fusion is joining two small nuclei.
2. The plutonium-239 nucleus absorbs a neutron and splits into two parts and a few extra neutrons. These neutrons are absorbed by more plutonium-239 nuclei which split – and so on.
3. To absorb neutrons and stop them causing more nuclei to fission. The rods are lowered or raised to change the number of neutrons absorbed and control the rate of reaction.
4. Stored under water in cooling tanks (high level waste).
5. $0.1 \text{ g} = 0.0001 \text{ kg} \times (3 \times 10^8 \text{ m/s})^2 = 9 \times 10^{12} \text{ J}$
6. The strong force.

Sample GCSE questions

1 This diagram shows a source of beta radiation used to control the thickness of paper produced by a factory manufacturing paper.

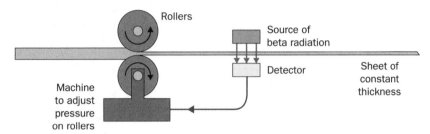

(a) What is beta radiation? **[2]**

Beta radiation is fast moving electrons emitted by radioactive nuclei when they decay.

There are two marks so you need to say more than 'electrons'

(b) Explain how the thickness of the paper is controlled. *The quality of your written communication will be assessed in this answer.* **[6]**

The beta radiation is directed from the source through the paper to the detector. Some of the radiation will be stopped by the paper. If the paper is too thin the amount of beta radiation passing through the sheet increases so a signal is sent to the rollers to decrease the pressure and make the sheet thicker. If the paper is to thick the amount of radiation reaching the detector will decrease so a signal is sent to the rollers to increase the pressure and make the sheet thinner.

Marks will be awarded depending on the number of relevant points included in the answer and the spelling, punctuation and grammar. In this question there are about 6 relevant points so 5 or 6 points with good spelling, punctuation and grammar will gain full marks.

(c) Why is alpha radiation or gamma radiation not used? **[2]**

Alpha radiation would all be absorbed by a thin sheet of paper and none would reach the detector. Gamma radiation is very penetrating and would all pass through a thick sheet of paper and reach the detector.

(d) The radioactive source used is strontium-90. Complete this equation for the radioactive decay of strontium-90 to the element yttrium. **[3]**

$$^{90}_{38}\text{Sr} \rightarrow \ ^{90}_{39}\text{Y} + \ ^{0}_{-1}\text{e}$$

[Total = 13]

Notice that you don't have to know anything about yttrium to complete the equation. The beta particle is represented as 'e'. Fill in mass number '0' and proton number of '−1' (because it has a charge of −1 and a proton is changed to a neutron when a beta particle is emitted) . To balance the equation the numbers for yttrium are 90 − 0 = 90 and 38 − (−1) = 39.

Sample GCSE questions

2 Radon-222 is a radioactive gas. This graph shows the radioactive decay of a sample of radon-222.

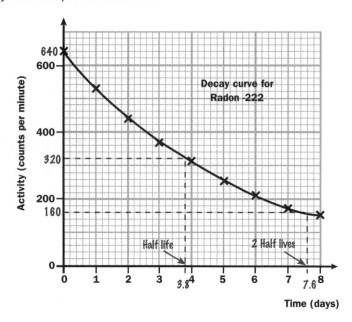

(a) Use the graph to find the half-life of radon-222 **[2]**

Half-life of radon-222 = ...**3.8 days**......

> Show on the graph that the half-life is the time for the count rate (640 per minute) to fall to half (320 per minute) Using two half-lives gives a bigger value and when halved the error will be smaller, so in this case it is a good idea to use two half-lives to work out your answer.

Radon-222 decays to polonium-218 by emitting an alpha particle.

(b) Complete this equation for the decay. **[3]**

$$^{222}_{86}Rn \rightarrow {}^{218}_{84}Po + {}^{4}_{2}He$$

> Fill in 4 and 2 for He.
>
> The mass number of polonium, 218, is given in the question, or can be worked out from 222 – 4 = 218. The number of protons is 86 – 2 = 84.

Radon gas is a decay product from radioactive rocks in the Earth's crust. It contributes to about 50% of the background radiation in the UK.

(c) Explain what is meant by 'background radiation'. **[1]**

Background radiation is the low level of radiation from the radioactive decay of sources that are all around us all the time.

(d) In some parts of the country radon gas can build up in houses. Explain why this is a health hazard. *The quality of your written communication will be assessed in this answer.* **[4]**

The radon gas breathed into the lungs will decay by emitting alpha particles. Alpha particles are very ionising and when they irradiate the lung tissue the cells may be damaged or killed. If the DNA is damaged the cells may mutate and become cancerous.

> Marks will be awarded depending on the number of relevant points included in the answer and the spelling, punctuation and grammar. In this question there are about 4 or 5 relevant points so 4 or 5 points with good spelling, punctuation and grammar will gain full marks.

[Total = 10]

Exam practice questions

1 **(a)** A patient is injected with Technetium-99 m which has a half-life of 6 hours. What fraction of the Technetium-99m nuclei are left after one day? **[1]**

....

(b) Why is it better to use an isotope with a half-life of 6 hours rather than:

(i) an isotope with a half-life of 6 minutes? **[1]**

....

(ii) an isotope with a half-life of 6 days? **[1]**

....

[Total = 3]

2 Read this information about radiation treatment for cancer.

To treat cancer of the thyroid gland patients are given a dose of iodine-131. Iodine is absorbed by the thyroid gland, so the radioactive iodine kills the cancer cells. After the treatment patients are radioactive for a few days. When the level drops to the normal background level patients can leave the hospital.

(a) Jack says he thought radiation caused cancer so how can radioactive iodine cure it? Explain how radiation can be both the cause and cure of cancer. **[3]**

....

....

....

(b) Sam has been treated for thyroid cancer. He says 'The benefit outweighed the risk'. Explain what he meant. **[2]**

....

....

Jo is a smoker. She avoided meeting Sam when he left hospital after being treated with a radioactive isotope because she was afraid Sam might be radioactive. Sam was not allowed to leave until his radioactivity level was below 3 mSv. He showed Jo the table below.

This table compares the risks of different activities by working out an average of days of life lost for each activity. This is worked out by:

total days of life lost by all the people who died early
_____
total population

Exam practice questions

Activity	Average days/years lost
Smoking 20 cigarettes a day	6 years
All accidents	207 days
Cancer due to being exposed to 3 mSv of radiation	15 days
Cancer due to being exposed to 10 mSv of radiation	51 days

Sam said that Jo was more at risk from smoking than from any remaining radioactivity.

(c) Explain why Sam was correct. [2]

....

....

....

(d) Suggest why Jo was more worried about radioactivity than smoking. [1]

....

[Total = 8]

3 Factory waste is discharged from a pipe at sea. To check that it is not being washed up on a nearby beach, a radioactive tracer is added. A suitable isotope to use would be: [1]

A an alpha emitter with a half-life of 2 days

B a beta emitter with a half-life of 2 days

C a beta emitter with a half-life of 1 year

D a gamma emitter with a half-life of 1 year.

☐

4 Put these statements in order to describe how the age of a wooden spear is determined using carbon dating. The first has been done for you. [5]

A A sample of the wood is tested to find the proportion of carbon-14.

B The proportion of carbon-14 in the wood falls as the nuclei decay.

C Carbon dioxide, containing some carbon-14, is taken in by the living tree during photosynthesis.

D The tree is cut down and no more carbon-14 is taken in.

E The result is compared with living wood and only half of the carbon-14 is left.

F The wood is made into a spear.

G The age of the spear is one half-life, which is 5730 years.

C						

Exam practice questions

5 Match the descriptions with the type of nuclear waste by drawing one straight line from each box on the left to one box on the right. **[2]**

Description	Type of waste
Used fuel rods from nuclear reactors	High level waste
Protective clothing worn by workers	Intermediate level waste
Reactor fuel containers	Low level waste

6 (a) What is nuclear fission? **[1]**

....

(b) How does the energy produced by nuclear fission compare with that produced in a chemical reaction? **[1]**

....

(c) Write **T** for the **true** and **F** for the **false** statements below. **[5]**

(i) Nuclear reactors use fuel rods made of uranium-235 or plutonium-239. ☐

(ii) A chain reaction occurs when a nucleus splits and releases a few neutrons which can be absorbed by other nuclei and cause them to split. ☐

(iii) Control rods are lowered into the nuclear reactor core to speed up the reaction. ☐

(iv) The energy released when the nuclei split heats up the reactor core. ☐

[Total = 7]

7 Scientists are researching the process of nuclear fusion with the aim of producing a fusion reactor. Explain the nuclear fusion process. *The quality of your written communication will be assessed in this answer.* **[6]**

....
....
....
....
....
....

8 Light

The following topics are covered in this chapter:

- Dispersion and total internal reflection
- Lenses
- Seeing images
- Telescopes and astronomy

8.1 Dispersion and total internal reflection

LEARNING SUMMARY

After studying this section, you should be able to:

- Describe and draw ray diagrams for refraction.
- Calculate speed from refractive index.
- Describe dispersion.
- Explain the conditions for total internal reflection.
- Calculate critical angle or refractive index.

Refractive index

AQA	P3	✓
OCR A	P7	✓
OCR B	P5	✓
EDEXCEL	P3	✓
CCEA	P2	✓

Refraction occurs when light changes speed (see page 36). The wavelength changes, but the frequency of the light stays the same.

Refraction in a glass block.

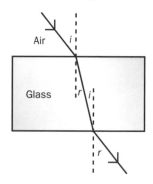

> **KEY POINT**
>
> The **refractive index**, n, of a medium:
>
> $$n = \frac{\text{speed of light in a vacuum}}{\text{speed of light in the medium}}$$

The refractive index is a ratio, so it has no units.

Snell's law states that the **angle of refraction**, r, when light enters a medium depends on the angle of incidence, i and the refractive index.

$$n = \frac{\sin i}{\sin r}$$

Lenses use refraction to change the direction of light rays.

Example: Find the speed of light, *v*, in a glass block.

Speed of light in a vacuum = 3×10^8 m/s. (The speed of light in air is so close to this that you can use this value, unless you are told otherwise.)

angle *i* = 30° angle *r* = 19°

$$n = \frac{\sin 30°}{\sin 19} = 1.54$$

$$\frac{3 \times 10^8 \text{ m/s}}{v} = 1.54$$

$$v = 1.95 \times 10^8 \text{ m/s}$$

Dispersion

OCR A	P7	✓
OCR B	P5	✓
CCEA	P2	✓

A prism separates white light into its different colours.

Dispersion by a glass prism.

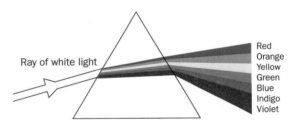

Ray of white light

Red
Orange
Yellow
Green
Blue
Indigo
Violet

> **Remember which colour bends the most by the mnemonic: Violet Veers Violently.**

Dispersion happens because the speed of light in the glass depends on the colour (the frequency) of the light. When white light passes from air to glass, violet rays are refracted towards the normal more than red rays. When the light passes from glass to air the violet rays are refracted away from the normal more than red rays.

In a prism this results in the white light being spread into a spectrum, as each colour from red through to violet is refracted more than the last.

Dispersion happens because all electromagnetic waves travel at the same speed in a vacuum, but in other media the speed of light depends slightly on the frequency of the radiation. Another way of saying this is that the refractive index depends on the frequency of the radiation. The higher the frequency, the greater the change in speed and refractive index.

Total Internal Reflection

AQA	P3	✓
OCR B	P5	✓
EDEXCEL	P3	✓
WJEC	P3	✓
CCEA	P2	✓

Total Internal Reflection (TIR) occurs when the angle of refraction is greater than 90° and the light cannot leave the medium. It only happens when light passes from a dense to a less dense medium, for example from glass to air, which means the angle of refraction is larger than the angle of incidence.

KEY POINT

The **critical angle** is the angle of incidence above which TIR occurs. It is the angle of incidence that gives an angle of refraction of 90° (see page 36).

The critical angle, c, of a medium with refractive index, n, is calculated using the equation:

$$\sin c = \frac{1}{n}$$

The higher the refractive index, the lower the critical angle. Diamond has very high refractive index and so a low critical angle which leads to more internal reflections. This is why diamonds sparkle so much.

Applications of TIR include:

Total internal reflection in prisms.

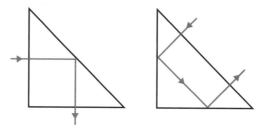

Reflecting prisms are used in:

- Binoculars.
- Cats-eyes in the road to reflect headlights.
- Cycle reflectors.

Optical fibres are used in:

- Communications (see page 57).
- Endoscopes.

An **endoscope** is two sets of **optical fibres** used for viewing inside the body or for performing minor surgical operations. Optical fibres are flexible so the endoscope can be passed into the body through an opening, for example the mouth or nose. The first set of fibres transmits light from an external light source into the body and the second set transmits the image out of the body.

Endoscopes can have tools attached for minor surgery. The optical fibres can also transmit a laser light beam, which can be used to cut and heat tissue to stop bleeding.

PROGRESS CHECK

1. Light is refracted away from the normal when it leaves water. What happens to its speed?
2. Which colour light travels slowest in glass?
3. Give one use of TIR.
4. Why does an endoscope need two bundles of optical fibres?
5. Calculate the refractive index of diamond that has $i = 20°$, $r = 8°$
6. Calculate the critical angle for glass ($n = 1.5$).

8.2 Lenses

LEARNING SUMMARY

After studying this section, you should be able to:

- Describe the effects of converging lenses on rays of light.
- Use the relationships between power and focal length.
- Calculate magnification.
- Draw ray diagrams for converging lenses.

Converging lenses

AQA	P3	✓
OCR A	P7	✓
OCR B	P5	✓
EDEXCEL	P1, P3	✓
CCEA	P2	✓

KEY POINT

A **converging lens** makes a **parallel** beam of light rays converge to a point, called the **focus**, **focal point** or **principal focus**. All converging lenses are fatter in the middle than at the edge.

KEY POINT

The **focal length**, f, of a lens is the distance from the centre of the lens to the focal point.

Converging lens with different focal lengths.

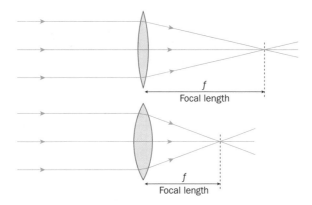

f
Focal length

f
Focal length

These lenses are also called **convex lenses**. When two lenses are made of the same material, the one with the greatest curvature will have the shortest focal length.

Light from a source spreads out, or **diverges**. A converging lens can be used to reduce the divergence and make a parallel or converging beam.

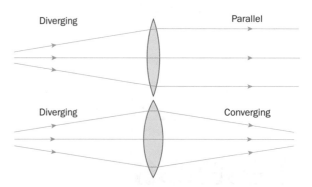

Using converging lenses to make a parallel or converging beam.

Power and magnification

AQA	P3	✓
OCR A	P7	✓
OCR B	P5	✓
EDEXCEL	P3	✓

KEY POINT

Power is measured in **dioptres** (D) where 1 D = 1 m^{-1}. The shorter the focal length, the more powerful the lens.

$$\text{Power (D)} = \frac{1}{\text{focal length (m)}}.$$

Examples:

- If a lens has a focal length of 0.8 m it has a power of 1 ÷ 0.8 = 1.25 D.
- A lens with a focal length of 10 cm has a power of 1 ÷ 0.1 = 10 D.
- Converging lenses have positive powers (for example +2 D) diverging lenses have negative powers (for example −0.5 D).

KEY POINT

The **magnification** produced by a lens is:

$$\text{Magnification} = \frac{\text{The image height}}{\text{The object height}}$$

Magnification has no units.

Images

AQA	P3	✓
OCR A	P7	✓
OCR B	P5	✓
EDEXCEL	P3	✓
CCEA	P2	✓

A lens is used to produce an **image**. The image is a copy of the **object**.

Image	Meaning
Real	It can be displayed on a screen because the light passes through it.
Virtual	It can only be seen by looking through the lens – the light doesn't pass through it.
Upright	The same way up as the object.
Inverted	Upside down compared to the object.
Magnified	Bigger than the object.

Remember that a real image is one you can touch, a virtual image not really there, like the image in a mirror or seen through a magnifying glass.

A converging lens can form real, inverted images, or virtual, upright images depending how far it is from the object.

Drawing ray diagrams

AQA	P3	✓
OCR A	P7	✓
OCR B	P5	✓
CCEA	P2	✓

You can find out what an image is like, and where it appears, by drawing a ray diagram using these facts:

- Light rays pass straight through the optical centre of the lens.
- Light rays parallel to the axis pass through the focal point (or light rays passing through the focal point leave the lens parallel to the axis).

What to do:

- Choose the scale.
- Draw a vertical line to represent the lens and a horizontal line to represent the principal axis through the centre of the lens.
- Mark the focal point in the right place on each side of the lens.
- Draw the object – an upright arrow the right size and in the right place.
- Draw a ray from the top of the arrow straight through the optical centre of the lens.
- Draw a ray from the top of the arrow parallel to the axis to the lens, and then through the focal point.
- Draw the image arrow, from the point where the rays meet to the axis.

> Ray diagrams are best drawn on graph paper with a sharp pencil and a transparent ruler (so you can see the whole diagram).

Ray diagrams.

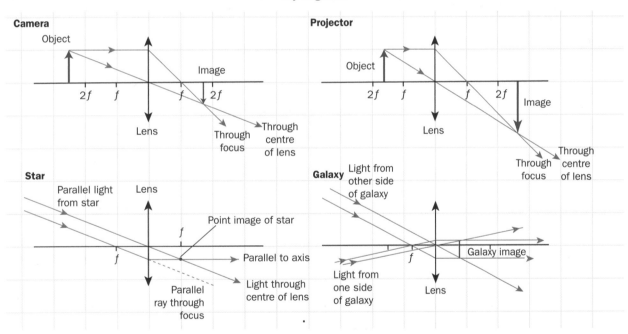

> **PROGRESS CHECK**
>
> 1. Lens A has a focal length of 10 cm and lens B has a focal length of 15 cm. Which lens is the fatter?
> 2. A converging lens brings light to a focus, but the focal point is too far away from the lens. Should a fatter or thinner lens be used?
> 3. Use the words from the image table on page 170 to describe the image in the ray diagram for the camera and the projector.
> 4. What is the focal length of a lens with power = 4 D?
> 5. The object measures 2.4 cm and the image measures 6 cm. What is the magnification?

8.3 Seeing images

LEARNING SUMMARY

After studying this section, you should be able to:

- Label a diagram of the eye.
- Explain short and long sight and how they can be corrected.
- Describe how a converging lens is used in a projector.
- Draw diagrams for diverging lenses.
- Describe the image formed by a magnifying glass.

The eye

AQA   P3   ✓
EDEXCEL P3   ✓

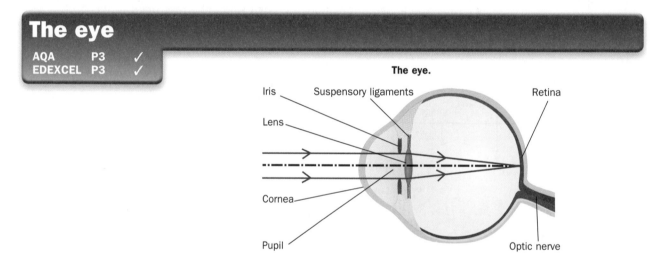

The eye.

Iris   Suspensory ligaments   Retina
Lens
Cornea
Pupil   Optic nerve

In the eye the convex lens and the cornea focus an image on the light sensitive cells of the **retina**. From these cells, signals are sent along the **optic nerve** to the brain.

The amount of light entering the eye depends on the size of the **pupil**, which is controlled by the **iris**.

The **cornea** is transparent and curved to refract light. The **ciliary muscles** alter the shape of the variable **lens**, so that the eye can focus on near and distant objects.

The average adult human eye can focus on objects between about 25 cm (the **near point**) and infinity from which rays are parallel (the **far point**).

Eyesight

AQA   P3   ✓
EDEXCEL P3   ✓

In short-sight, you can see things a short distance away. For far objects, to stop the eye focussing in such a short distance needs a diverging lens.

In long-sight, you can see things a long way away. To stop the eye needing such a long distance to focus needs a converging lens.

KEY POINT

Spectacles, or contact lenses, are worn to correct:

- **Short-sight** where near objects are seen clearly, but distant objects are blurred. Corrected with a **diverging lens**.
- **Long-sight** where distant objects are seen clearly, but near objects are blurred, or viewing them causes eye strain. Corrected with a **converging lens**.

Short-sightedness is caused by the eyeball being too long, or the lens being unable to focus, so the image is in front of the retina.

Long-sightedness is caused by the eyeball being too short, or the lens being unable to focus, so the image is behind the retina.

Lenses for correcting eyesight.

Diverging lens

Short sight

and its correction

Light from a distant object

Converging lens

Long sight

and its correction

Light from a near object

The **focal length** of the lens depends on its refractive index and its curvature, so special glass and plastics with high refractive index are useful for making flatter lenses which are thinner.

Laser surgery is an alternative way of correcting eyesight. A laser is used to alter the curvature of the cornea so that the light rays are focussed on the retina.

Uses of lenses

AQA   P3   ✓
OCR A   P7   ✓
OCR B   P5   ✓
CCEA   P2   ✓

The converging lens in a camera produces a real, inverted image that is smaller than the object. The lens must be more than twice the focal length from the object so that it produces a small image, (see page 171).

If you are asked to compare the eye and the camera don't make statements only about the eye or only about the camera.

In many ways the camera is similar to the eye (see pages 48–49). However, to focus the image the camera lens is moved:

- Away from the film to focus on a close object.
- Towards the film to focus on a distant object.

The converging lens in the **projector** produces a real, inverted image that is magnified. The lens must be between one and two times the focal length from the slide or film so that it produces a magnified image. To focus the image the projector lens is moved:

- Away from the slide to focus on a closer screen and give a smaller image.
- Towards the slide to focus on a distant screen and give a more magnified image.

Projector.

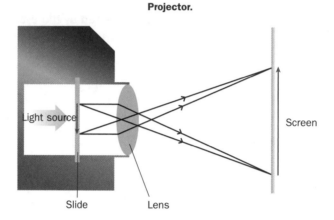

When a converging lens is used as a **magnifying glass** it is closer to the object than the focal length. The image is virtual, upright, and magnified. You must look through the lens to see it. The image appears to be further away than the object.

Ray diagram for a magnifying glass.

Diverging lenses

AQA	P3	✓
EDEXCEL	P3	✓
CCEA	P2	✓

A diverging lens makes a parallel beam diverge so that it appears to have come from the focus. Diverging lenses produce virtual images that are upright and smaller than the object.

A diverging lens.

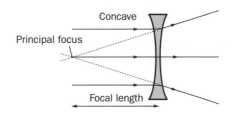

Ray diagram for a diverging lens.

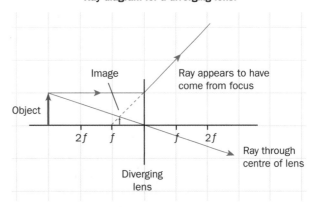

8.4 Telescopes and astronomy

LEARNING SUMMARY

After studying this section, you should be able to:

- Select lenses for the eyepiece and objective of a telescope.
- Calculate the magnificaton of a telescope.
- Compare a refracting and reflecting telescope.
- Describe how angles of parallax are used to calculate distances.
- Explain and use the parsec as a unit of distance.

Simple telescopes

OCR A   P7   ✓
EDEXCEL   P1   ✓

> **KEY POINT**
>
> A simple optical telescope has two converging lenses:
>
> - The eyepiece is the lens nearest your eye.
> - The objective is the lens nearest the object.

The objective produces a real image. The eyepiece is used as a magnifying glass to magnify this image.

The eyepiece is the more powerful lens – the fatter, or more curved lens.

> **KEY POINT**
>
> $$\text{Magnification} = \frac{\text{focal length of objective lens}}{\text{focal length of eyepiece lens}}$$

Example: A telescope has an eyepiece with power 2 D and an objective with power 0.25 D.

Focal lengths are $f_e = \dfrac{1}{2} = 0.5$ m and

$f_o = \dfrac{1}{0.25} = 4$ m

Magnification $= \dfrac{f_o}{f_e} = \dfrac{4 \text{ m}}{0.5 \text{ m}} = 8$

The telescope has an **angular magnification** of ×8. Looking through this telescope makes the Moon *appear* 8 times (or ×8) bigger or 8 times nearer than without the telescope.

The Moon through a telescope.

Stars are so far away they still look like points through a telescope, but groups of stars are spread out.

Aperture and brightness

OCR A   P7   ✓

The objective of the telescope gathers all the light rays that enter it and focuses them to a point. This makes the star brighter. You can see stars that are too dim to see with the naked eye. The diameter of the objective is called the **aperture**. To collect radiation from weak or distant sources you need a telescope with a large aperture.

Diffraction

OCR A   P7   ✓

The radiation entering the telescope is diffracted (see page 52) by the aperture. To reduce diffraction the aperture must be much larger than the wavelength of the radiation otherwise the telescope will not give sharp images.

Reflecting telescopes

OCR A   P7   ✓
EDEXCEL P1   ✓

Most astronomical telescopes have a **concave mirror**, not a converging lens as the **objective**. There are lots of different designs, but all have a very large mirror to focus all the light gathered. The eyepiece lens is then positioned to look at the focussed light. The mirror is concave, curved so that parallel rays are focussed at a single point.

A concave reflector for a telescope

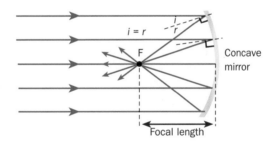

> A concave mirror has a caved-in shape.

Advantages of using a mirror:

- A lens focuses different colours (frequencies) at slightly different points, so you do not get such a clear image as with a mirror.
- You can make very large mirrors and support them from behind, but you can't make such large lenses because:
  - the lens would be so heavy it would distort under its own weight
  - it would be difficult to make the glass even.

Life elsewhere in the Universe

OCR A   P7   ✓

Astronomers have identified hundreds of distant stars with planets. They have not discovered alien life, existing now or in the past. However, because there are so many stars, scientists think it likely that life has evolved somewhere else in the Universe. The Search for Extra-Terrestrial Intelligence (SETI) project scans radio waves from space for patterns that suggest aliens using radio waves for communication.

Parallax and parsecs

OCR A   P7   ✓

Parallax was introduced on page 76. Astronomers measure the angle they turn their telescopes in six months to observe a star and record the **angle of parallax**.

The angle of parallax for a close and distant star.

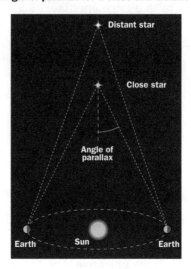

> ### KEY POINT
> The angle of parallax is half of the angle moved against a background of stars in six months. The smaller the angle, the further away the star.

Astronomers measure distance with a unit called the **parsec** (pc).

> ### KEY POINT
> One parsec is the distance to a star with a parallax angle of one second of arc.

- 1/60th of a degree (1°) is 1 minute of arc (1').
- 1/60th of a minute is 1 second of arc (1").

A **parsec** (3.1×10^{16} m) is similar to a **light year** (9.5×10^{15} m):

- Interstellar distances are typically a few parsecs.
- Intergalactic distances are typically a few megaparsecs (Mpc).

Get a scientific calculator. They are useful for calculations with large numbers. Make sure you know how to use it and take it to the exam.

The size of the Universe

OCR A   P7   ✓

How bright a star is depends on the radiation it emits. This is called its **luminosity** and is a measure of how bright it really is. How bright it appears to us depends on how far away it is and is called its **brightness**.

Cepheid variable stars can be used to measure **intergalactic distances**. These stars pulse in brightness. This graph shows the **variation in brightness** over several days. The period of the variation is about 5.3 days.

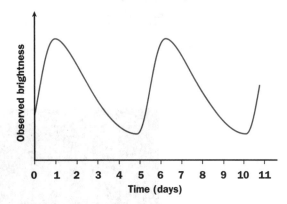

The variation in brightness of the star Delta Cephei

For OCR A you also need to know about the apparent movements of stars, planets and the Moon, including retrograde motion and eclipses, and how they are caused. You need to know that astronomical positions are described by two angles.

The astronomer Henrietta Leavitt discovered a number of Cepheid variable stars in a nearby galaxy. Because they were all the same distance from Earth, she realised that the brighter the Cepheid, the longer the period of variation in brightness. She plotted a graph of luminosity against the period of variation of brightness. This gave us a way of working out the luminosity from the period of variation – and we can measure the period from Earth.

When astronomers discover a new galaxy they can work out how far away it is by:

- Finding a Cepheid variable star in the galaxy and measuring its observed brightness.
- Working out its period of variation of brightness.
- Using the graph of luminosity against period of variant of brightness to find its luminosity.
- Using the luminosity and average brightness seen on Earth to work out the distance to the star (and the galaxy).

Edwin Hubble used Cepheid variable stars to measure distances, see page 80. They were very important in determining the size of the Universe, and showing that most nebulae were distant galaxies.

Hubble also looked at spectra of stars and discovered the red-shift. There are two ways to produce a spectrum. One is a **prism** and the other is a **diffraction grating.**

For OCRA you also need to know about energy levels in atoms and absorption and emission of photons.

Hubble's Law relates the speed at which a galaxy is moving away from us to its distance away.

> **KEY POINT**
>
> Hubble's Law can be written as an equation:
>
> **Speed of recession (km/s) = Hubble constant (km/s per Mpc)**
> **× distance (Mpc)**

Data from Cepheid variable stars in distant galaxies have been used to calculate the **Hubble constant**.

Stars and the Hertzsprung-Russell diagram

OCR A   P7   ✓

The Hertzsprung-Russell diagram is a graph of the luminosity of stars against their temperature.

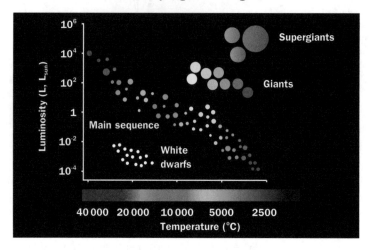

Make sure you know the regions where each type of star is found.

The **core** of a star is where most nuclear fusion takes place. In a main sequence star, hydrogen is fused to helium in the core. Energy is transported from the core to the surface by convection and by photons of radiation. The star's surface is called the **photosphere**. Energy is radiated into space from the photosphere.

Nucleosynthesis of elements

WJEC   P3   ✓

The Big Bang model suggests that before any stars formed the Universe was about 75% hydrogen and 25% helium with very small amounts of other light elements. Fred Hoyle and other scientists used the results of nuclear research to account for the formation of all other elements by **nucleosynthesis** in stars.

PROGRESS CHECK

1. Jenna makes a telescope from two lenses. Lens A = 1.5 D and B = 1.0 D. Which is the eyepiece and which is the objective?
2. Which is the best choice of telescope objective, lens diameter 2 cm or 5 cm?
3. Describe the objective of a reflecting telescope.
4. Have scientists detected signs of life on Mars?
5. Calculate the angular magnification of Jenna's telescope in Q.1.
6. Two stars have parallax angles 1" and 0.5". Which is closer?

6. The one with angle 1".
5. 1.5
4. No
3. A large concave mirror.
2. 5 cm – will collect more light.
1. eyepiece = A objective = B

Sample GCSE questions

1 Look at this diagram of the human eye.

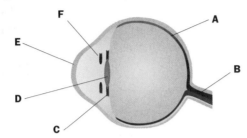

(a) Choose the letter which points to:

   (i) The cornea. ....E.. **[1]**

   (ii) The retina. ....A................................. **[1]**

(b) What is D? ....the lens............................... **[1]**

(c) Complete this diagram to show how a perfect eye focuses the light from a distant point. **[1]**

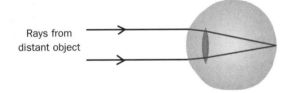

Rays from distant object

(d) Megan can see near objects but cannot focus on objects in the distance and has been prescribed spectacles. Write an explanation for Megan of how her eyes are different to the eye shown above, and how the lenses correct her vision. You can include a labelled diagram or diagrams in your answer. *The quality of your written communication will be assessed in this answer.* **[6]**

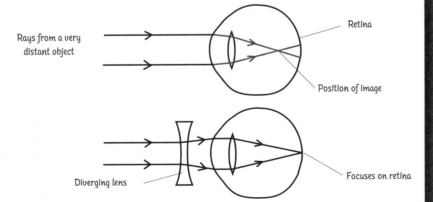

Rays from a very distant object · Retina · Position of image · Diverging lens · Focuses on retina

Megan's eyes focus the rays of light in front of the retina. She is given diverging lenses to make the parallel rays diverge so that they are brought to a focus on the retina.

[Total = 10]

Learn all of the points labelled in the diagram and make sure you can spell them correctly.

There is some focussing by the front of the eye, and more by the lens. The important point is that it should focus on the retina. Make sure you can draw how the rays are focussed for perfect, near (or short)-sighted and far (or long)-sighted eyes.

Marks will be awarded depending on the number of relevant points included in the answer and the spelling, punctuation and grammar. In this question there are about 6 relevant points so 5 or 6 points with good spelling, punctuation and grammar will gain full marks. The marks will be awarded if shown on a labelled diagram or stated, by doing both you are less likely to miss any.

Sample GCSE questions

2 These diagrams show rays of light travelling from glass to air as the angle of incidence, *i* is increased. Diagram C is incomplete.

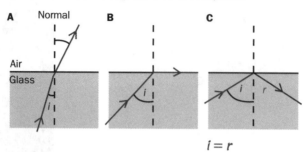

$i = r$

(a) Explain why the ray bends as shown in diagram A. **[2]**

The speed of light increases as the light travels from glass to air, so the ray is refracted away from the normal.

(b) What name is given to the value of the angle of incidence, *i* in B? **[1]**

The critical angle

(c) Complete diagram C to show the path of the ray. **[2]**

(d) Describe an endoscope and explain how it works. *The quality of your written communication will be assessed in this answer.* **[6]**

An endoscope is a flexible tube that can be used for seeing inside objects and into otherwise inaccessible places. They are used, for example, by doctors to look inside the body. They have two bundles of optical fibres inside, one set for transmitting light from a source to illuminate the object, and the other for carrying the light rays reflected from the object back to form an image. The light travels along the fibres and total internal reflection keeps the light rays inside the fibres.

[Total = 11]

There are 2 marks so explain that the speed changes and this is refraction and give the direction of the change, it increases so refraction is away from the normal.

There are 2 marks for this part, so show that the ray is reflected and make sure *i = r*

Marks will be awarded depending on the number of relevant points included in the answer and the spelling, punctuation and grammar. In this question there are about 7 relevant points so 5 or 6 points with good spelling, punctuation and grammar will gain full marks. You should include what the endoscope does, what it is made of, and the physics of how it works.

Exam practice questions

1 A magnifying glass is used to look at a crystal that is 3 mm wide. The magnification is × 4. How wide is the image? **[1]**

A 3 mm

B 4 mm

C 7 mm

D 12 mm

☐

2 Jack makes a telescope using two lenses.

Lens A has a power of +1.25 D and a diameter of 4 cm.

Lens B has a power of + 2.5 D and a diameter of 5 cm.

Tick (✓) true or false for each statement. **[6]**

Statement	True	False
Lens A should be used as the eyepiece lens and lens B as the objective lens.		
The focal length of lens B is 40 cm.		
The magnification of the telescope is × 2.		
With the naked eye, the Moon has an angular size of about 0.5°. With this telescope the Moon has an angular size of about 2°.		
Through the telescope the Moon appears upside down (inverted).		
Increasing the aperture of the telescope will make the image of the Moon brighter.		

3 Rays of light from a distant object are brought to a focus by a converging lens as shown.

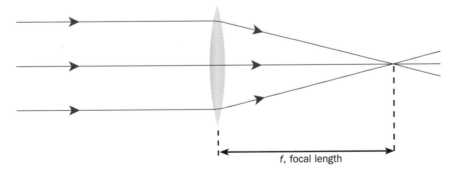

f, focal length

(a) If the lens was an eye lens, what change would happen to bring light rays from a **near** object to a focus at the same point? **[1]**

....

(b) If the lens was a camera lens, what change would be needed to bring light rays from a **near** object to a focus at the same point? **[1]**

....

[Total = 2]

Exam practice questions

④ Which of these statements about the image formed by the lens in a camera is correct? **[1]**

A It is real, smaller than the object and inverted.

B It is real, smaller than the object and the right way up.

C It is virtual, smaller than the object and inverted.

D It is virtual, smaller than the object and the right way up.

☐

⑤ A star moves 1 second of arc against the fixed stars in 6 months. **[1]**

(a) What is the angle of parallax?

....

(b) Explain why it is 2 parsecs away. **[2]**

....

....

[Total = 4]

⑥ This diagram shows an object O placed between the principal focus F and a converging lens.

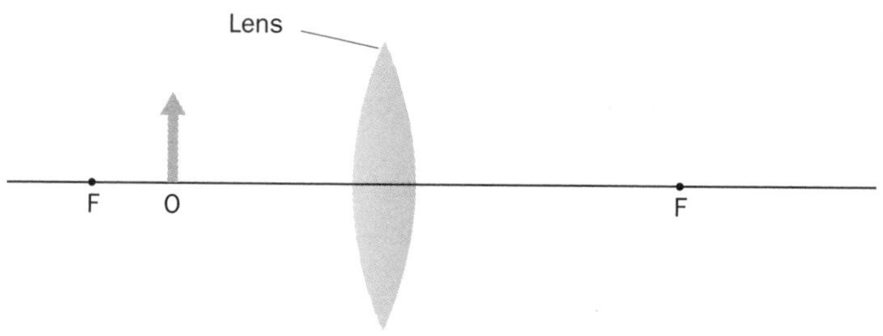

(a) Draw two rays of light to complete the diagram to show the image of O formed by the lens. **[4]**

(b) The lens is held near a window and forms an image of a distant house on a screen.

State three properties of the image on the screen. **[3]**

1. ....

2. ....

3. ....

[Total = 7]

Exam practice questions

7 Alex is making a telescope. This diagram shows rays of light from a very distant object falling on a converging lens.

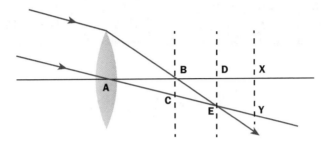

(a) Which letter shows the position of the image of the distant object? .... **[1]**

(b) What distance (between which two letters) is the focal length of the lens? .... **[1]**

(c) Alex finds another lens with the same focal length but a larger diameter.
What difference will this make to the image of the very distant object? **[1]**

....

(d) This table shows the lenses Alex has to choose from.

Lens	Focal length (cm)	Diameter (cm)
A	50	10
B	25	12.5
C	2.5	7.5
D	5	5

(i) Which lens has the highest power? .... **[1]**

(ii) Which lens, used as the objective lens, would give the brightest image? .... **[1]**

(iii) Alex decides to use lenses A and C. Explain which lens should be the eyepiece,
which the objective, and calculate the magnification of the telescope. **[3]**

....

....

....

(iv) Shannon uses lenses C and B. Describe and explain the differences Alex and
Shannon would notice between their telescopes. **[4]**

....

....

....

[Total = 12]

9 Further physics

The following topics are included in this chapter:

- **Electromagnetic effects**
- **Kinetic theory**
- **The gas laws**
- **Medical physics**
- **More forces and motion**

9.1 Electromagnetic effects

LEARNING SUMMARY

After studying this section, you should be able to:

- Describe how a motor works.
- Explain how to increase the force turning the motor.
- Describe how a generator works.
- Explain how to increase the induced voltage.
- Explain how transformers work and calculate voltages using the turns ratio.

Electromagnets

AQA	P3	✓
OCR B	P6	✓
WJEC	P3	✓

When an electric current flows in a conductor a **magnetic field** is set up around it. This diagram shows the shape of the magnetic field around a straight wire, a coil, and a long coil called a **solenoid**. The solenoid is an **electromagnet**. It will attract and repel other magnets. The advantage of a solenoid compared to a bar magnet is that it can be switched on and off. They have many uses, e.g. a crane lifts and drops scrap metal using an electromagnet.

The magnetic field around a current carrying wire, coil and solenoid

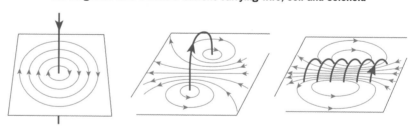

The motor

AQA	P3	✓
OCR A	P5	✓
OCR B	P6	✓
WJEC	P3	✓
CCEA	P2	✓

There is a force on a current carrying conductor when it is at right angles to a **magnetic field**.

The force is:

- At right angles to both the current and the magnetic field.
- Reversed if the current or the magnetic field is reversed.

The motor effect.

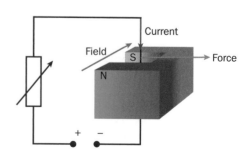

- Increased if the current or the magnetic field is increased.
- Zero if the current is parallel to the magnetic field.

When a horizontal loop of wire is placed in a magnetic field the current in one side of the loop produces a force upwards and on the other side it produces a force downwards.

A d. c. motor

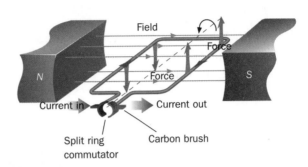

This pair of forces makes the loop rotate to a vertical position. The force is made larger by using a coil with more turns of wire.

The commutator reverses the current direction as the coil passes through the vertical position. This keeps the coil rotating in the same direction.

Uses of motors

AQA	P3	✓
OCR A	P5	✓
OCR B	P6	✓
WJEC	P3	✓
CCEA	P2	✓

Motors are widely used. Examples are in:

- Domestic appliances like washing machines and electric drills.
- Electric motor vehicles.
- Hard disk drives.
- DVD and CD players.

The force and motion direction is given by the **left hand rule**.

Left hand rule.

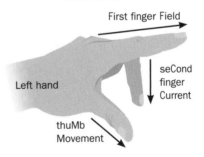

The magnets have curved pole pieces to decrease the size of the gap and increase the magnetic field strength.

Some motors have electromagnets instead of permanent magnets, while some have a rotating magnet and fixed coils.

Electromagnetic induction

AQA	P3	✓
OCR A	P5	✓
OCR B	P6	✓
WJEC	P3	✓
CCEA	P2	✓

KEY POINT

Electromagnetic induction occurs when a changing magnetic field **induces** a voltage in a conductor.

It is used in generators and in transformers. If there is an induced voltage in a coil and the ends are connected to make a complete circuit then a current flows in the coil.

A generator

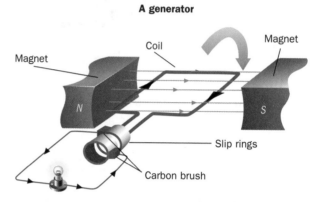

As the coil is rotated voltage is induced in the coil. The output is a.c.

The induced voltage can be increased by:

- Increasing the speed of rotation of the magnet or electromagnet.
- Increasing the strength of the magnetic field.
- Increasing the number of turns on the coil.
- Placing an iron core in the coil.

The direction of the current is given by the **right hand rule**.

Right hand rule.

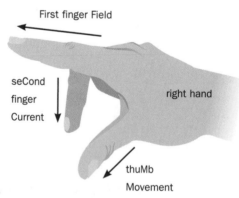

If you use the wrong hand rule you will get the wrong direction. Many married people wear a ring on the left hand. Remember: M for married = motor and motion produced.

The word induced describes the voltage that results from movement and magnetic fields. Always use it for generators, never for motors.

This graph shows how the size of the induced voltage alters as the coil rotates.

a.c. voltage and rotation.

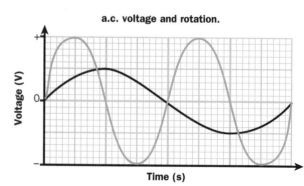

When the coil rotates twice as fast, in other words at twice the frequency, the frequency of the a.c. doubles and the voltage increases. This is shown by the blue line on the graph.

Transformers

AQA	P3	✓
OCR A	P5	✓
OCR B	P6	✓
WJEC	P3	✓
CCEA	P2	✓

A transformer changes the size of an alternating voltage.

A transformer is made from a primary coil and a secondary coil wound on an iron core to concentrate the magnetic field in the coils.

A transformer.

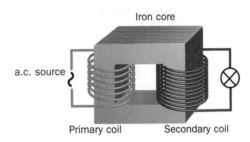

An alternating current in the primary coil produces a changing magnetic field. The changing magnetic field induces a voltage in the secondary coil.

> For AQA and WJEC you need to be able to apply the equation $P = IV$ to the primary and secondary coils.

Step-up transformers increase the size of the output voltage by having more turns in the secondary coil than the primary coil. Step-down transformers reduce the size of the output voltage by having fewer turns on the secondary coil than the primary coil.

KEY POINT

The ratio of the voltages in the coils is:

$$\frac{\text{primary voltage}}{\text{secondary voltage}} = \frac{\text{number of turns in primary coil}}{\text{number of turns in secondary coil}}$$

$$\frac{V_P}{V_S} = \frac{N_P}{N_S}$$

Switch Mode Transformers

AQA	P3	✓

The current flows in the primary coil of a transformer even when there is nothing connected to the secondary coil (when there is no **load** connected to the secondary coil), so energy is wasted heating up the coil and the iron core. If the frequency of the a.c. is increased so that the magnetic field is changing more rapidly the efficiency of the transformer is improved.

Switch mode transformers operate at a high frequency often between 50 kHz and 200 kHz a.c. Mains has a frequency of only 50Hz. So switch mode transformers have circuits inside, which switch the current on and off very rapidly to increase the frequency. They also use very little power when no load is connected to the secondary coil. Less energy is wasted and the transformer can be lighter and smaller.

Switch mode transformers are useful for mobile phone chargers and other similar devices, because they are smaller, lighter and waste less energy, especially if they are left plugged in when they are not being used.

PROGRESS CHECK

1. Which direction does the motor in the diagram on page 187 rotate?
2. How can a motor be made to rotate faster?
3. What would happen to the generator on page 188 if the magnetic field was reversed?
4. What effect does a step-up transformer have on the voltage?
   A transformer has input voltage 230 V and 920 turns on the primary coil.
5. How many turns are needed on the secondary to give an output voltage of 12 V?
6. If there are 96 turns on the secondary what is the output voltage?

6. 230 V × 96 ÷ 920 = 24 V
5. 920 × 12 ÷ 230 V = 48
4. It increases it.
3. The current would be reversed.
2. Increase current, magnetic field strength, number of turns in coil.
1. Anti-clockwise.

9.2 Kinetic theory

LEARNING SUMMARY

After studying this section, you should be able to:

- Use the kinetic theory to describe solids, liquids and gases.
- Use the kinetic theory to explain pressure and temperature.
- Explain and calculate latent heat.
- Use the kelvin scale of temperature.

Solids, liquids and gases

AQA	P1	✓
OCR B	P1	✓
EDEXCEL	P3	✓
WJEC	P3	✓
CCEA	P1	✓

The **kinetic theory** says that matter is made of particles:

- **Solid** objects are held together by forces between the particles and have a regular shape.
- **Liquid** particles have enough energy to break the inter-molecular bonds and slide over each other.
- **Gas** particles have enough energy to separate completely.

A sketch of particles must show that the particle size doesn't change. Solid particles are regularly spaced and touching.

The liquid particles are still touching. There are no gaps large enough for another particle to fit in.

Gas particles are very widely spaced, so do not draw too many.

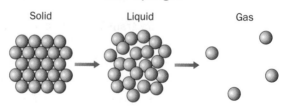

Solid–liquid–gas.

Solid   Liquid   Gas

The particles vibrate more as the object's temperature increases. The **temperature** increases when the kinetic energy of the particles increases.

Latent heat

OCR B   P1   ✓

Most of the time, heating an object raises its temperature. However, at the melting point or boiling point of a substance it will absorb heat energy without getting any hotter. This heat energy is required to change the state of the substance and it is known as **latent heat**. All changes of state happen at a constant temperature.

For OCR B you need to be able to use the equation $E = mL$ where L is specific latent heat.

- Solids melt when the particles have enough energy to break the bonds holding them together.
- Liquids freeze when the particles lose this energy to the surroundings and form stronger bonds with touching particles.
- Liquids boil when the particles have enough energy to separate completely. This happens to particles anywhere in the liquid.
- Gases condense when the particles lose energy and form weak bonds with other particles.
- Liquids evaporate from the surface when the most energetic particles have enough energy to leave the liquid and become a gas. Evaporation happens at any temperature above melting point.

Pressure

AQA   P3   ✓
OCR A   P7   ✓
OCR B   P5   ✓
EDEXCEL   P3   ✓
WJEC   P3   ✓

When a force acts on a small area it exerts a greater pressure than when it is spread over a large area. Pressure depends on area as well as force. Pressure is measured in pascals (Pa).

$$\text{pressure (Pa)} = \frac{\text{force (N)}}{\text{area (m}^2)}$$

$1\ Pa = 1\ N/m^2$

A gas is made up of a huge number of atoms or molecules that are constantly moving. The particles collide randomly with other particles and the walls of the container. The forces exerted by all the particles as they collide with inside walls of the container add up to give the gas pressure. The pressure is equal in all directions.

Reducing the volume of a gas increases the pressure.

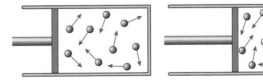

A common mistake is to say that gas pressure results from collisions between particles.

The movement of the particles in the gas is random. They change speed and direction, as they collide with other particles and the walls of the container. This means they change momentum. The force exerted is the rate of change of momentum. The total of all the forces exerted by the particles per square metre of the container walls is the gas pressure.

The kelvin temperature scale

OCR A	P7	✓
EDEXCEL	P3	✓
WJEC	P3	✓

KEY POINT

Absolute zero is the temperature at which all particles stop moving. They have no kinetic energy. On the celsius scale absolute zero is –273°C. The **kelvin temperature scale** is a scale that starts at absolute zero. The unit is the kelvin (K). One kelvin is the same size as one Celsius degree.

Temperatures in kelvin are 273 bigger than those in degrees Celsius, because they start from the lowest possible temperature and they cannot be negative.

Comparing the kelvin and Celsius temperature scales.

To convert between temperature in kelvin and degrees Celsius:

- Temperature in K = temperature in °C + 273.
- Temperature in °C = temperature in K – 273.

The advantage of using the kelvin temperature scale is that it is an absolute scale starting at zero. This means doubling the temperature doubles the average kinetic energy of the particles. (This is not true if the temperature is measured in °C.) The temperature is directly proportional to the average kinetic energy of the particles.

PROGRESS CHECK

1. What happens to the heat energy absorbed by ice at 0°C?
2. A force of 4000 N acts on an area of 2 m². What is the pressure?
3. What is 100°C in kelvin?
4. What is 195 K in °C?
5. The temperatures of gas A is 100 K and gas B is 200 K. What does this tell you about the average kinetic energy of their particles?

1. The heat energy is used to break the bonds between the ice molecules as it melts. This happens at constant temperature.
2. 4000 N ÷ 2 m² = 2000 Pa
3. 100 + 273 = 373K
4. 195 – 273 = –78°C
5. Particles of gas B have twice the average KE of particles of gas A.

9.3 The gas laws

LEARNING SUMMARY

After studying this section, you should be able to:

- Describe the relationship between pressure, temperature and volume for a fixed mass of gas.
- Use of relationships PV = constant, $\frac{P}{T}$ = constant and $\frac{V}{T}$ = constant.
- Use the gas law $\frac{PV}{T}$ = constant.

A fixed mass of gas

OCR A	P7	✓
EDEXCEL	P3	✓
WJEC	P3	✓

The gas in a sealed container has mass, measured in kilograms (kg), and volume, measured in metres cubed (m³). The mass is fixed, but by squeezing or heating it you can change the volume. The pressure is measured in pascals (Pa) and the temperature is measured in kelvin (K).

Pressure and temperature

OCR A	P7	✓
OCR B	P5	✓
EDEXCEL	P3	✓
WJEC	P3	✓

When the container has a constant volume and is heated the temperature increases and the pressure increases.

Heating a gas with constant volume.

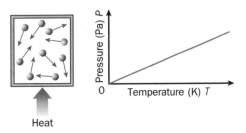

Heat

> **KEY POINT**
>
> Doubling the temperature doubles the pressure. The graph is a straight line through (0, 0).
>
> The pressure is proportional to the temperature.
>
> $$\frac{\text{pressure (Pa)}}{\text{temperature (K)}} = \text{constant}$$

When the gas is heated its temperature rises. The particles have more energy so their average speed is faster.

- A faster particle exerts more force during a collision.
- A faster particle will make more collisions.

These changes increase the pressure.

Volume and temperature

OCR A	P7	✓
EDEXCEL	P3	✓
WJEC	P3	✓

When the gas is heated at a constant pressure the temperature increases and the volume also increases to keep the pressure constant.

Heating a gas at constant pressure.

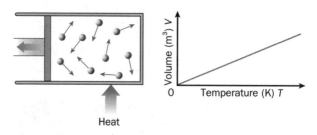

> ### KEY POINT
>
> Doubling the temperature doubles the volume. The graph is a straight line through (0, 0).
>
> The volume is proportional to the temperature
>
> $$\frac{\text{volume (m}^3)}{\text{temperature (K)}} = \text{constant}$$

The forces exerted by the particles push the sides of the container out to create a larger volume.

Pressure and volume

OCR A	P7	✓
OCR B	P5	✓
EDEXCEL	P3	✓
WJEC	P3	✓

When the gas is compressed at a constant temperature the volume decreases and the pressure increases.

Compressing a gas at constant temperature.

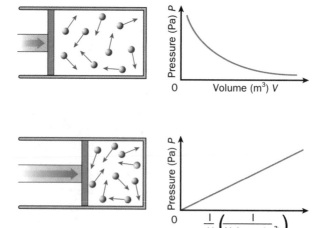

Halving the volume doubles the pressure. The graph of P against $\frac{1}{V}$ is a straight line through (0, 0).

The pressure is inversely proportional to the temperature.

pressure × volume (m^3) = constant

The particles have the same kinetic energy so if the container is larger the forces exerted by the particles on the sides of the container will be less.

Using the equations

OCR A	P7	✓
EDEXCEL	P3	✓
WJEC	P3	✓

P is pressure, T is temperature and V is volume. At constant volume:

$$\frac{P_1}{T_1} = \frac{P_2}{T_2}$$

> You must convert temperatures from degrees Celsius to kelvin. Otherwise these equations will not work, because zero pressure and volume occur at –273°C.

Example: Pressure increases from 1×10^5 Pa to 2×10^5 Pa. The temperature was 27°C. Calculate the new temperature T_2.

$T_1 = 273 + 27 = 300$ K

At constant pressure:

$$\frac{V_1}{T_1} = \frac{V_2}{T_2}$$

At constant temperature:

$$P_1 V_1 = P_2 V_2$$

An ideal gas

OCR A	P7	✓
EDEXCEL	P3	✓
WJEC	P3	✓

These relationships show that if you cool an ideal gas to absolute zero the particles:

- Stop moving completely and have no kinetic energy.
- Make no collisions and exert no pressure.
- Take up no space and have no volume.

A real gas does not behave like this. It condenses to a liquid before reaching absolute zero. These equations work for gases that are above their boiling points, for example nitrogen at room temperature, but not water vapour.

The gas law

OCR A	P7	✓
EDEXCEL	P3	✓
WJEC	P3	✓

We will now collect together the relationships for pressure temperature and volume.

For a constant mass of gas, with pressure P, temperature T (in kelvin) and volume V:

$$\frac{P_1 V_1}{T_1} = \frac{P_2 V_2}{T_2}$$

Example: A medical oxygen cylinder contains 0.004 m³ of compressed oxygen at a pressure of 1.5×10^7 Pa and has a temperature of 15°C. What volume of oxygen will this give when released from the cylinder?

Atmospheric pressure = 1.01×10^5 Pa and temperature = 20°C.

$V_1 = 0.004$ m³

$V_2 = ?$

$P_1 = 1.5 \times 10^7$ Pa

$P_2 = 1.01 \times 10^5$ Pa

$T_1 = (273 + 15)$ K = 288 K

$T_2 = (273 + 20)$ K = 293 K

$$\frac{1.5 \times 10^7 \text{ Pa} \times 0.004 \text{ m}^3}{288 \text{ K}} = \frac{1.01 \times 10^5 \text{ Pa} \times V_2}{293 \text{ K}}$$

$V_2 = 0.604$ m³

PROGRESS CHECK

For a fixed mass of ideal gas what happens to:

1. The pressure when you double the volume (temperature is constant)?
2. The volume, at one third the pressure (temperature is constant)?
3. The temperature in K, when volume is halved (pressure is constant)?
4. The particles at absolute zero?
5. Tyre pressure at 30°C is 6×10^5 Pa. What is the pressure at 21°C?
6. $P_1 = 1 \times 10^5$ Pa, $V_1 = 0.5$ m³, $T_1 = 290$ K, $P_2 = 1.2 \times 10^5$ Pa, $V_2 = 0.8$ m³. Calculate T_2.

1. Pressure halves
2. Volume is three times greater
3. Temperature in K is halved
4. They stop moving, have no kinetic energy
5. $T_1 = 30 + 273 = 303$ K $T_2 = 21 + 273 = 294$ K $P_2 = \frac{6 \times 10^5 \text{ Pa} \times 294 \text{ K}}{303 \text{ K}} = 5.8 \times 10^5$ Pa
6. $T_2 = \frac{1.2 \times 10^5 \text{ Pa} \times 0.8 \text{ m}^3 \times 290 \text{ K}}{1 \times 10^5 \text{ Pa} \times 0.5 \text{ m}^3} = 557$ K

9.4 Medical physics

LEARNING SUMMARY

After studying this section, you should be able to:

- Describe ultrasound.
- Explain how an ultrasound scan works.
- Calculate distances directly or from oscilloscope traces.
- Describe Positron Emission Tomography (PET).

Ultrasound

AQA	P3	✓
OCR B	P4	✓
EDEXCEL	P1, P3	✓
CCEA	P2	✓

Ultrasound is sound with a frequency above the 20 kHz upper threshold of human hearing. Ultrasound is produced by electronic systems and by animals, for example dogs and bats.

Ultrasound waves are partially reflected at a boundary between two different materials. The rest of the wave is transmitted. A pulse of ultrasound is transmitted and the returning pulse is detected. The time for a reflection to return is recorded and the distance to the boundary calculated.

Examples are:

- Sonar, to locate the sea floor and shoals of fish.
- Medical diagnosis, for example scans to get information about the eye or the fetus.

Ultrasound scanning of the fetus works like this:

- An ultrasound pulse is emitted from the probe.
- The same probe detects the reflected pulse.
- The two pulses are displayed on an oscilloscope or a computer is used to process all the reflections and produce an image.

Ultrasound.

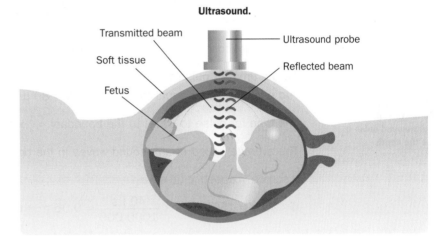

- The time for the pulse to travel from the emitter to the fetus and back is recorded.
- The speed of the ultrasound waves is 1500 m/s. (speed = distance/time)
- The depth, d, is half the distance the wave travelled:

$$\text{depth} = \frac{\text{speed} \times \text{time}}{2}$$

When ultrasound scans moving objects there is a change in frequency called a Doppler shift. This effect can be used to measure the speed of blood flow in the body.

Ultrasound can also be used to treat patients, for example to break down kidney stones or gallstones. The vibrations caused by the ultrasound break down the stones.

Advantages over X-ray scans are that ultrasound scans:

- They do not damage living cells or DNA because they are not ionizing.
- They can produce images of soft tissue.

X-rays need to be used if a bone image is required.

Write these bullet points on flash cards. Use them to help you learn the uses of ultrasound and the steps involved in measuring depth.

Ultrasound scans

OCR B   P4   ✓
EDEXCEL   P1, P3   ✓
CCEA   P2   ✓

Example:

Measuring distance with ultrasound.

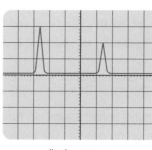

display screen

timebase control

The diagram shows the distance between the pulses on the display is 4.0 divisions.

The oscilloscope is set to 20 μs per division.

So the time for pulse to travel to the boundary and back is the time between the two peaks which is equal to 4.0 × 20 μs = 80 μs.

The time for pulse to travel to the boundary = $\frac{1}{2}$ × 80 μs = 40 μs.

The average speed of ultrasound waves in the body is 1500 m/s.

distance = speed × time

distance = 1500 m/s × $\dfrac{40\ \mu s}{1000\ 000}$ = 0.06 m = 6 cm

> Remember that the distance travelled by the ultrasound pulse is *twice* the depth.

Positron emission tomography (PET)

EDEXCEL   P3   ✓

The **positron** is the antiparticle of the electron. It has the same mass and equal, but opposite charge.

When it meets an electron they annihilate each other. The mass is converted to energy which is transferred by gamma rays.

To conserve momentum, two equal gamma rays are produced that travel in opposite directions.

An electron and a positron annihilate each other.

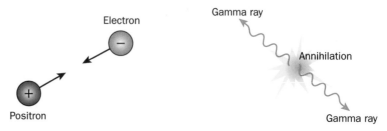

In **positron emission tomography (PET)** the patient is injected with an isotope that decays by emitting positrons and placed in a scanner that detects gamma rays.

The isotope decays.

* The positrons travel less than a millimetre before positron-electron annihilation.

- The two gamma rays travel in opposite directions and are detected. Only pairs of gamma rays are used in analysis, any other rays are ignored.
- A picture is built up of the positron concentration.

The isotope is chosen to be appropriate for its use. New isotopes with short half-lives can be produced by bombarding stable isotopes with protons shortly before being used.

To look at cells that use glucose, like brain cells or tumour cells, a nucleus of fluorine-18 is used to replace an oxygen nucleus in a glucose molecule. After injection the glucose spreads through the body. The body is then scanned.

β+ Decay

The positron is also called a β⁺ particle. It is represented in nuclear equations as $_{1}^{0}e$

β+ decay: a **proton** in an unstable nucleus changes into a **neutron** and emits a **positron**.

Fusion of hydrogen to helium in stars releases positrons:

$$4{}_{1}^{1}H \rightarrow {}_{2}^{4}He + 2{}_{1}^{0}e$$

PROGRESS CHECK

1. What is ultrasound?
2. Sonar speed = 1500 m/s, time for pulse to return = 0.08 s. Calculate the water depth.
3. Which particles from protons, positrons and electrons have the same **a)** charge and **b)** mass?
4. What would be the depth of a boundary which produced pulses five divisions apart on the oscilloscope screen on page 198?
5. Write a nuclear equation for the decay of a proton to a neutron.

5. ${}_{1}^{1}H \rightarrow {}_{0}^{1}n + {}_{1}^{0}e$
   divisions = 6 cm so 5 × 4 ÷ 6 = 5 × 4 ÷ 6 = 7.5 cm)
4. time = 5 × 20 μs, time ÷ 2 = 50 μs d = 1500 m/s × 50 μs = 0.075 m = 7.5 cm (or 4
3. a) proton, positron b) positron, electron
2. d = ½ × 1500 m/s × 0.08 s = 60 m
1. sound with frequency greater than 20 kHz

9.5 More forces and motion

After studying this section, you should be able to:

LEARNING SUMMARY

- Calculate moments and use the principle of moments.
- Explain how to find the centre of mass of an object.
- Describe how the period of a pendulum depends on its length.
- Describe how levers and hydraulic systems magnify the effect of forces.
- Explain how a force keeps on object moving in a circle.

Moments

AQA   P3   ✓
CCEA   P2   ✓

When a force acts on an object that is free to **rotate** about a **pivot** (a fixed point), the force has a turning effect, which is called a **moment**. The size of the **moment** depends on:

- The size of the force.
- The distance of the point where it is applied from the pivot.
- The angle at which the force is applied.

> **KEY POINT**
>
> The **moment** (or turning effect) of a force is worked out from
>
> **moment (Nm) = force (N) × perpendicular distance to the pivot (m)**
>
> $M = F \times d$
>
> The moment of a force is measured in Nm.

> If the line of action of the force passes through the pivot, there is no turning effect.

The moment can be **clockwise** or **anticlockwise**.

> **KEY POINT**
>
> **The principle of moments:**
>
> When a system is balanced, about any pivot:
>
> the sum of the anticlockwise moments = the sum of the clockwise moments.

The clockwise moments equal the anticlockwise moments.

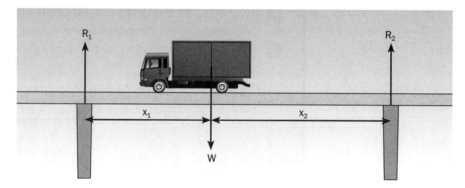

Taking moments about an axis through the left hand support:

$R_2 \times (x_1 + x_2) = W \times x_1$

Or, taking moments about an axis through the right hand support:

$R_1 \times (x_1 + x_2) = W \times x_2$

Notice that as the lorry moves across the bridge x_1 gets smaller and x_2 gets bigger. At the same time R_1 gets bigger and R_2 gets smaller so that the equations above are still true.

Centre of mass

The **centre of mass** of an object is the point at which all the mass of an object seems to be concentrated. It is the point where the weight appears to act. For a human body it is just behind the 'tummy button.' The point does not have to be inside the object – for a metal ring it is in the centre of the ring.

The centre of mass of objects

> For CCEA centre of mass is called centre of gravity.

If in object is suspended freely it will stop swinging and hang so that the centre of mass is directly below the point of suspension.

Finding the centre of mass of a piece of card.

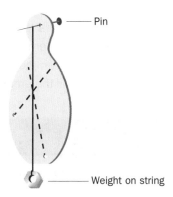

Pin

Weight on string

The centre of mass of a thin sheet of material can be found by:

- Suspending the shape from a pin in a loose-fitting whole so that it is free to swing.
- Using a plumb-line to mark the vertical line from the pinhole.
- Repeating this at other points.

All the lines will cross at the same point, and this point is the centre of mass. You can check this by turning the shape to a horizontal position and balancing it on your finger. The only place it will balance is when your finger is at the centre of mass.

For a symmetrical object, the centre of mass is on the axis of symmetry.

Stability

AQA	P3	✓
CCEA	P2	✓

If the centre of mass of an object moves so that the line of action of the weight is outside the base of the object there will be a moment on the body that will make it topple over. In the diagram, the traffic cone has a moment that will make it fall back onto its base, so it is **stable**, whereas the bowling pin has a moment that will tip it over, so it is **unstable**.

A stable traffic cone and unstable bowling pin.

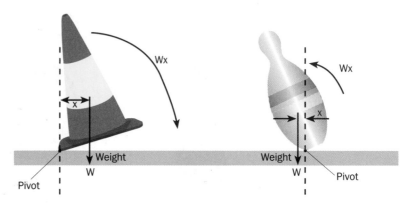

The traffic cone is more stable than the bowling pin because:

- The traffic cone has a wider base.
- The traffic cone has a lower centre of mass.

An object may have a low centre of mass because of its shape, like the traffic cone, or because there is more mass near the base. Double-decker buses are more stable than they look because the extra mass of the engine and transmission near the ground gives them a low centre of mass.

Levers

AQA	P3	✓

A **lever** is a simple machine that can be thought of as a force multiplier.

For the lever in the diagram, If the clockwise moment (Effort × d_E) is greater than the anticlockwise moment (Load × d_L) then the load will be lifted up.

A simple lever.

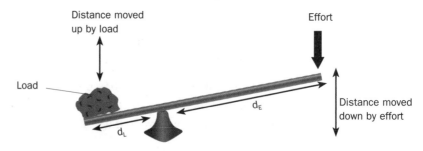

The longer the lever to the right of the pivot, i.e. the length d_E, the smaller the effort needed to lift the load.

The work done is the same because the effort has to be moved down much further than the load is lifted up, and the work done is the force x distance moved in the direction of the force.

The pendulum

AQA   P3   ✓

A simple **pendulum** is a small, heavy weight, or **bob**, on a string. When the bob is pulled to one side and released it will keep swinging. The motion can be described by:

- The **amplitude** of the swing – how far the bob is pulled to one side.
- The **length**, l – the length of the string.
- The **period**, T – the time from one complete swing – from release back to your hand.
- The **frequency**, f – the number of swings per second.

$$\text{period (s)} = \frac{1}{\text{frequency (Hz)}}$$

$$T = \frac{1}{f}$$

The weight does not affect the period and nor does amplitude.

> **KEY POINT**
>
> The period of a pendulum depends on the length, the longer the pendulum the longer the period.

A child's swing behaves like a pendulum.

Hydraulics

AQA   P3   ✓

Unlike gases liquids are almost incompressible. When you squeeze a plastic bag full of liquid:

- the pressure is transmitted throughout the liquid
- the pressure acts equally in all directions.

> **KEY POINT**
>
> $$\text{pressure (Pa)} = \frac{\text{force (N)}}{\text{area (m}^2)}$$
>
> $$P = \frac{F}{A}$$

The same pressure can be the result of a large force on a large area, or a small force on a small area. This is made use of in hydraulic systems. By using different cross-section areas on the effort and load side of a hydraulic system, the system can act as a force multiplier. In the diagram the force is multiplied by three because the area of the piston is one third the size of the other.

Car brakes and power steering systems use hydraulics, as does construction machinery like bulldozers and cranes.

A hydraulic system.

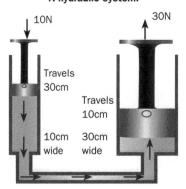

Circular motion

| AQA | P3 | ✓ |
| CCEA | P2 | ✓ |

When an object moves in a circular path it is constantly changing direction. Its velocity is changing so it is accelerating, even if its speed is constant.

> Acceleration is change of velocity. Velocity is speed *and* direction.

There must be a resultant force on the object to make it accelerate (change direction.) This force is called a **centripetal force**, it changes the direction of the object, not its speed.

> **KEY POINT**
>
> For an object to move in a circle there must be a resultant force on it towards the centre of the circle. This force is called the **centripetal force**.

When an object on a string moving in a circle. The centripetal force is provided by the tension in the string. Cut the string and the object flies off at a tangent to the circle – a straight line.

A circular path needs a resultant force towards the centre of the circle.

When a car goes round a circular track the centripetal force is provided by the friction between the tyres and the road.

For astronomical objects that are in orbit, the centripetal force is provided by gravity, for example planets orbiting the Sun (or the Moon and artificial satellites orbiting the Earth).

To make an object move in a circle, the centripetal force needed increases as:

- the mass of the object increases
- the speed of the object increases
- the radius of the circle decreases.

PROGRESS CHECK

1. On one side of a see-saw is a 7 N force 2 m from the pivot, on the other side is a 2 N force 3 m from the pivot and a 4 N force 2 m from the pivot. Does the see-saw balance?

2. When you stand on your right foot, you keep your balance by leaning towards the right. Explain why.

3. A load of 10 N is 40 cm from the pivot. If the maximum effort is 2 N, how far must the effort be from the pivot?

4. A hydraulic jack lifts a weight of 100 N, the piston area is 50 cm^2.
   a) What effort is needed on the piston of 8 cm^2?
   b) The weight is lifted 40 cm. How far has the effort piston moved?

5. On a fairground ride the cars move in a circle.
   a) In what direction is the force on a rider to keep them moving in a circle?
   b) What happens to the force when the car speeds up?
   c) What happens to the force when the cars move in to travel in a circle of smaller radius?
   d) Is the force larger on an adult or a small child?

1. Yes, 7 N × 2 m =14 Nm and 2 N × 3 m + 4 N × 2 m = 14 Nm
2. Leaning moves your centre of mass, which is behind your 'tummy button' over the base formed by your right foot. If it is outside the base you will topple over.
3. 10 N × 40 cm ÷ 2 N = 200 cm = 2 m.
4. a) 100 N ÷ 50 cm^2 × 8 cm^2 = 16 N b) 50 cm^2 × 40 cm ÷ 8 cm^2 = 250 cm.
5. a) towards the centre of the circle b) it increases c) it increases d) on the adult.

Sample GCSE questions

1 Louise is travelling on an aircraft. She drinks the water from a plastic bottle and puts the top back on the empty bottle. The air pressure in the cabin is 76 kPa.

(a) What causes the air molecules inside the bottle to exert a pressure?

The forces exerted by the air molecules as they collide with the inside plastic walls of the bottle all add up to give the pressure.

[2]

> Remember that it is not the molecules colliding with each other that causes gas pressure, but with the walls of the container.

(b) Atmospheric pressure at the destination airport is 100 kPa.

(i) What will happen to the plastic bottle? **[1]**

A It will be compressed. **C** It will explode.
B It will expand. **D** It will stay the same.

A

> There are 2 marks for this part, and 'because the pressure is higher outside' or 'lower inside' is not enough for both marks.

(ii) Explain why this happens. **[2]**

The pressure is higher outside the bottle than inside the bottle, so the bottle will be squashed until the pressures are equal and the air takes up a smaller volume.

(iii) The inside volume of the plastic bottle is 500 ml = 0.5×10^{-3} m³. Calculate the volume of the bottle after landing. (You can assume that the temperature stays the same and the air in the bottle is dry.) **[3]**

> Show your working. If you make a mistake in the final calculation you may still get a mark for the earlier part. You might prefer to write $p_1V_1 = p_2V_2$ as the equation you are using.

$PV = $ constant so $76 \text{ kPa} \times 0.5 \times 10^{-3} \text{ m}^3 = 100 \text{ kPa} \times V$

$V = \dfrac{76 \text{ kPa} \times 0.5 \times 10^{-3} \text{ m}^3}{100 \text{ kPa}} = 0.38 \times 10^{-3} \text{ m}^3$

Volume of bottle = ..0.38 x 10⁻³.. m³

[Total = 8]

> A step-down transformer is used to reduce voltage. The primary coil has more turns than the secondary coil.

2 This diagram shows a transformer.

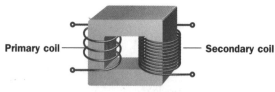

Primary coil —   — Secondary coil

Explain how a voltage across the primary coil in transformer results in a voltage across the secondary coil. *The quality of your written communication will be assessed in this answer.* **[6]**

The alternating voltage across the primary coil causes an alternating current in the primary coil. This sets up a magnetic field which is continually changing. The changing magnetic field in the iron core cuts the turns of the secondary coil and induces a voltage in the secondary coil.

> Marks will be awarded depending on the number of relevant points included in the answer and the spelling, punctuation and grammar. In this question there are about 6 relevant points so 5 or 6 points with good spelling, punctuation and grammar will gain full marks.

Exam practice questions

1 **(a)** Explain how the construction of a step-up transformer is different to a step-down transformer. [1]

....

(b) Explain why mains electricity is a.c. rather than d.c. *The quality of your written communication will be assessed in this answer.* [6]

....

....

....

....

(c) Describe three ways that the output voltage of a generator can be increased. [3]

1. ....

2. ....

3. ....

[Total = 10]

2 Match the situation shown in the left hand boxes to the temperature in kelvin shown in the right hand boxes. [3]

Situation	Temperature
Body temperature 37°C	273 K
Absolute zero of temperature	0 K
Freezing temperature of water	236K
November in the Antarctic –37°C	310 K

3 Tick (✓) **true** or **false** for each statement. [5]

When the volume of a gas is reduced to half the size ...	True	False
its pressure doubles if the temperature is unchanged.		
its temperature (in °C) doubles if the pressure is unchanged.		
the temperature (in K) halves if the pressure is unchanged.		
the average kinetic energy of the particles increases if the pressure is unchanged.		
the pressure increases because the particles collide more often with each other.		

4 At what temperature do the particles of a gas have twice as much kinetic energy as at 20°C? [2]

....

....

Exam practice questions

5 **(a)** What is ultrasound? [1]

....

(b) Ultrasound is used to build up an image of the unborn fetus. Explain what happens to the ultrasound when it reaches the fetus. [3]

....

....

(c) State another medical use, other than scanning parts of the body, of ultrasound. [1]

....

[Total = 5]

6 A patient is given a tracer that is a gamma emitter and after a short time a gamma camera is used to produce a scan of the patient. A second patient is given a PET scan. PET stands for Positron Emission Tomography.

(a) Write down one similarity between the two techniques. [1]

....

(b) Write down one difference between the two techniques. [1]

....

(c) Explain how radioisotopes that are positron emitters can be produced artificially. [1]

....

(d) How is the positron emitter carried to the part of the body that is under investigation? [2]

....

(e) Give one example of when a PET scan is used. [1]

....

[Total = 6]

7 Ultrasound waves are used to scan the fetus in the womb.

(a) X-rays would give an image of the fetus. Explain why X-rays are not used. [1]

....

(b) The ultrasound travels through soft tissue and is reflected from bone.
The pulse takes 0.02 ms to return. [3]
What is the distance to the bone?
(The speed of ultrasound in soft tissue is 1550 m/s.)

....

....

[Total = 4]

Exam practice answers

Chapter 1

1. D
2. A
3. (a) carpet is an insulator (or stone is a better conductor)
   (b) the layer of air between is a good insulator
   (c) convection currents carry smoke upwards
   (d) hot water rises by convection, the cold water falls and is heated.
   (e) black surfaces radiate more heat and cool quicker.
4. (a) A = furnace B = boiler C = turbine D = generator
   E = transformer
   (b) uranium (or plutonium)
   (c) In a nuclear power station the energy from the nuclear reaction/fission is used to heat the water to steam to turn the turbines. In a coal-fired station coal is burned to heat the water.
   (d) Advantage e.g. large amount of energy from small amount of fuel or does not produce carbon dioxide. Disadvantage e.g. produces radioactive waste which must be stored safely for thousands of years, accident could cause release of radioactive material which could contaminate a wide area.
5. (a) the amount of wind (or wind speed)
   (b) C
   (c) (i) the appliances used at that time of day, e.g. cooker
       (ii) at night everything might be switched off
   (d) If the wind drops below the maximum the turbine will produce less; The turbine will not cope with higher than average demand.
   (e) To cope with peaks in demand, or can be used when there is no wind (or to give a more constant supply)
   (f) (i) 500 000   (ii) 1 500 000
   (g) The time for the turbine to save the amount of money that it cost to buy and install
6. C
7. (a) The maximum voltage will be greater.
   (b) The voltage = 0 V
   (c) The voltage will be about – 0.5 mV (or will be negative)
   (d) The voltage will be about – 0.5 mV (or will be negative)
8. (a) efficiency = (8W ÷ 100W) × 100% = 8%
   (b) It is transferred to heat.
   (c) 40% of electrical power, $P = 8$ W, $40P ÷ 100 = 8$ W,
       $P = (100 × 8$ W$) ÷ 40 = 20$ W
   (d) $3 × 100$W $× 7 = 300 × 7$ W h $= 0.3 × 7$ kWh $= 2.1$ kWh
9. Answer should describe advantages and disadvantages of nuclear and gas power stations and compare the two.

Chapter 2

1. (a) T (b) F (c) T (d) T (e) F (f) T (g) F
2. (a) 45 cm ÷ 5 = 9 cm
   (b) $v = f\lambda$ or wave speed = frequency × wavelength, 9 cm × 4 Hz = 36 cm/s
   (c) The waves will speed up
3. (a) The waves are refracted.
   (b) The waves are reflected.
   (c) They set off an explosion and the waves travel through the Earth (1 point) being reflected and refracted at boundaries between rock layers (1 point). They record where the waves arrive at the surface (1 point), and can work out the path of the waves (1 point) this tells them where the boundaries between the rock layers are (1 point).
   Marks will be awarded depending on the number of relevant points included in the answer and the spelling, punctuation and grammar. In this question there are 5 relevant points so 4 or 5 points with good spelling punctuation and grammar will gain full marks.
4. (a) Diffraction
   (b) The wave crests will be closer together (1 point) and the harbour entrance is now greater than the wavelength (1 point) so the diffraction effect will be less (1 point). More like (b) on page 37. A diagram must be labelled and clearly show these points if they are not written.
   Marks will be awarded depending on the number of relevant points included in the answer and the spelling, punctuation and grammar. In this question there are only 3 relevant points so all 3 points with good spelling punctuation and grammar will gain full marks.
5. (a) longitudinal
   (b) wavelength = wave speed ÷ frequency, wavelength = 330 m/s ÷ 256 Hz = 1.29 m
6. (a) A = crust B = mantle C = liquid outer core D = solid inner core
   (b) Tectonic plates
   (c) The plates float on the mantle (1 point) which behaves like a very thick liquid (1 point). Convection currents (1 point) in the mantle (1 point) mean that hot material rises (1 point) at some points and then moves away from these hot spots (1 point) until it cools and sinks (1 point). The tectonic plates on top of the mantle move away from the hot spots with the hot mantle material (1 point).
   Marks will be awarded depending on the number of relevant points included in the answer and the spelling, punctuation and grammar. In this question there are 8 relevant points so 6 to 8 points with good spelling punctuation and grammar will gain full marks.
7. P-waves arrive before the S-waves (1 point), but S-waves have larger amplitude and are more destructive (1 point). If warnings are given when P-waves arrive (1 point) then people are prepared for the S-waves (1 point) . They can move to safer places (1 point). Some equipment is designed to automatically shut down (1 point) when P-waves are detected. Example: (1 point) In Japan the high speed trains start to slow down and stop and nuclear reactors are shut down.
   Marks will be awarded depending on the number of relevant points included in the answer and the spelling, punctuation and grammar. In this question there are 8 relevant points so 6 to 8 points with good spelling punctuation and grammar will gain full marks.

Chapter 3

1. radio, infrared, visible light, ultraviolet
2. Radio waves are blocked by the ionosphere.
3. They both have a lens to focus the picture on the light sensitive material (retina or screen) at the back. The light sensitive screen produces an electric signal.
4. A beam of microwaves transmitted through the atmosphere to a receiver will be absorbed by the water droplets so that if the beam is reduced or blocked there is rain present.
5. D

Exam practice answers

6. Example similarities: Both heat food by causing molecules to vibrate. Both heat using electromagnetic radiation. Both rely on conduction to spread heat through the food and cook the inside. Example differences: grill uses infrared, microwave oven uses microwaves. Infrared vibrates molecules on surface, microwaves penetrate about 1 cm into the food and vibrate the water molecules.

7. The hills have gaps roughly the same size as the wavelength of the radio waves so they are diffracted – they spread out around the hills. The microwaves for phones have smaller wavelength and are not diffracted so the straight beams are blocked.

8. Example answer: The hole in the ozone layer was caused by gases called CFCs (1 point) which react with the ozone (1 point), leaving a gap in the ozone layer (1 point). The ozone layer protects life on Earth from the high energy ionising (1 point) ultraviolet radiation (1 point) from the Sun. The hole in the ozone layer leads to more ultraviolet radiation reaching the Earth's surface (1 point) and causing skin cancer and cataracts (1 point). The governments of the world have passed laws (1 point) banning CFCs and the ozone hole is starting to reduce in size (1 point).
Marks will be awarded depending on the number of relevant points included in the answer and the spelling, punctuation and grammar. In this question there are 9 relevant points so 8 to 9 points with good spelling, punctuation and grammar will gain full marks.

9. True: a, b, f. False: c,d,e.

Chapter 4

1. (a) The geocentric model is that the Earth is at the centre and the planets and stars move around it. The heliocentric is that the Sun is at the centre of the solar system and all the planets including Earth orbit around it in approximate circles.
   (b) because the idea conflicted with people's religious beliefs.

2. Two of: The Moon has less dense rocks and only a small iron core. The Earth and Moon rocks are the same. The Moon has igneous rocks but no sign of recent volcanic activity.

3. (a) (i) yes (b) (ii) yes (c) (iii) no
   (b) The advantages of unmanned missions are: The costs of unmanned missions are less than manned missions, because there is no need to take food, water, oxygen and other supplies people need. There are no lives at risk if something goes wrong with an unmanned mission, but there are with a manned mission. The disadvantages of an unmanned mission are that everything must be planned before an unmanned mission leaves – no adjustments, repairs or changes can be made unless they are programmed into the computers and can be done remotely. Also, most people are more interested in manned missions. Unmanned missions do not inspire people in the way that manned missions do.
   Marks will be awarded depending on the number of relevant points included in the answer and the spelling, punctuation and grammar. There must be advantages and disadvantages given for full marks.

4. (a) A light year is the distance travelled by light through space, or a vacuum, in a year.

(b) The Sun is much closer than the other stars, it is much less than 1 light year away.

5. Diameter of the Earth = 12 760 km, Diameter of the Solar System = 10 light years, Diameter of the Milky Way galaxy = 100 000 light years, Distance to the nearest star = 4 light years.

6. (a) A and D
   (b) All the spectral lines are shift towards the red end of the spectrum
   (c) It means that the star is moving away from us.

7. (a) The Big Bang theory is that the Universe started to expand from a small initial point about 14 thousand million years ago. It is still expanding. The evidence is that the electromagnetic radiation from all the distant galaxies is red–shifted, and the further away the galaxy is the more its radiation is red shifted. This shows that all the distant galaxies are moving away from us and those further away are moving fastest. Scientists predicted that there should be cosmic microwave background radiation left from the big bang, and this was found in the 1960s.
   Marks will be awarded depending on the number of relevant points included in the answer and the spelling, punctuation and grammar.
   (b) They should try to repeat their findings and other scientists should try to reproduce them. They should look for another explanation, but not discard the theory until they can work out a new one that accounts for all the data.
   (c) They cannot answer this question because they cannot collect data or find evidence to answer it. It has more to do with religious belief than cause and effect. It cannot be tested.

Chapter 5

1. (D) BCEAF (1 mark for B anywhere before C, 1 mark for C anywhere before E, etc)

2. (a) 20m/s × 1200 kg = 24 000 kg m/s
   (b) C
   (c) Total momentum before = 24 000 kg m/s + 45 000 kg m/s = 69 000 kgm/s
   = total momentum after = (1200 kg + 3000 kg) × velocity.
   Velocity = 69 000 kgm/s ÷ 4200 kg = 16.4 m/s

3. (a) 800 kg × 10 N/kg × 80 m = 640 000 J
   (b) gain in KE = loss in PE = 640 000 J
   (c) 640 000J = ½ mv^2 = ½ × 800kg v^2 , v = $\sqrt{}$ (1600 m/s) = 40 m/s
   (d) 25 000 J ÷ 10 m = 2500 N

4. (a) steady speed, resultant force = 0, so force = 5000 N
   (b) ½ × 2000 kg × (30 m/s)^2 = 900 000 J

5. (a) 12 + 24 = 36
   (b) If the driver is tired, or affected by some drugs (including alcohol and some medicines), or distracted and not concentrating the time for him or her to react will be increased, during this time the car travels further so the thinking distance is increased.
   If the road is wet or icy, or the tyres or brakes are in poor condition, the friction forces will be less so the braking distance is increased.
   If the vehicle is fully loaded with passengers or goods the extra mass reduces the deceleration during braking, so the braking distance is increased.

Exam practice answers

Marks will be awarded depending on the number of relevant points included in the answer and the spelling, punctuation and grammar. In this question there are 6 relevant points so 5 or 6 points with good spelling punctuation and grammar will gain full marks.

(c) (i) It stretches (ii) Force = change of momentum ÷ time (or rate of change of momentum) The momentum changes to zero when the dummy stops. The force on the dummy depends on the time this takes. If the time to stop is longer the force is smaller.

Chapter 6

1. (a) The fence is given a negative charge so that it attracts the positively charged paint drops.
   (b) By giving the paint drops and the fence opposite charges the paint droplets will be attracted to the fence and repelled from each other so they will spread out evenly over the metal surface and even curve round to coat the back of the fence as shown in the diagram. This means that less paint will be wasted by falling on the ground or drifting away in the air and the paint coat will be even, and get in to hard to reach corners.
   Marks will be awarded depending on the number of relevant points included in the answer and the spelling, punctuation and grammar. In this question there are about 6 relevant points so 5 or 6 points including two advantages will gain full marks.

2. Paddles are placed on the patient's chest. They must make a good electrical contact with the body. To make the heart muscle contract, electric charge is passed from one paddle to the other through the body. This starts the heart beating steadily again.
   Marks will be awarded depending on the number of relevant points included in the answer and the spelling, punctuation and grammar. In this question there are 3 or 4 relevant points so 3 or 4 points with good spelling punctuation and grammar will gain full marks.

3. (a) current = charge ÷ time = 300 C ÷ 60 s = 5 A
   (b) voltage = energy ÷ charge
       energy = 12 V × 300 C = 3600 J
   (c) $P = IV$ power = 5 A × 12 V = 60 W

4. (a) the case is a non-conductor
   (b) (i) current flows normally (ii) fuse melts and breaks the circuit.
   (c) the fuse will not melt until 13A flows, this is a much higher current than needed and damage will be done before the fuse melts.

5. (a) left hand switch = X, right hand switch = Y.
   (b) no
   (c) yes
   (d) You can switch on only the fan so that it can be used for cooling. You cannot switch on only the heater in order to prevent it from overheating. The air from the fan is needed to cool the heating element.
   (e) 230 V ÷ 70 Ω = 3.3 A
   (f) 3.3 A + 0.6 A = 3.9 A

6. (a) Light dependent resistor
   (b) 1000 Ω + 500 Ω = 1500 Ω
   (c) (i) (15 V ÷ 1500 Ω) × 1000 Ω = 10 V
       (ii) (15 V ÷ 1500 Ω) × 500 Ω = 5 V

(d) As the light level increases the resistance decreases so the p.d. across the LDR decreases. (Although question says changes, you need to say increase and decrease in the answer.)
Marks will be awarded depending on the number of relevant points included in the answer and the spelling, punctuation and grammar. In this question there are 3 steps so all 3 points with good spelling, punctuation and grammar will gain full marks.
(e) As the light level increases the drop in p.d. could switch a light off, (or as the light level decreases the increase in p.d. could switch a light on).

Chapter 7

1. (a) one sixteenth bi) It would decay away before reaching the organ and being recorded by the gamma camera bii) the patient would stay radioactive for days.
2. (a) The radiation is ionizing so it can kill or damage cells. It can be used to kill cancer cells. If it damages healthy cells they can turn cancerous.
   (b) The benefit of killing the thyroid cancer outweighed the risk of causing another cancer.
   (c) The risk of being exposed to 3 mSV of radiation amounts to 15 days lost on average, but the risk from smoking amounts to 6 years lost on average, so smoking is much more dangerous.
   (d) People are more worried about unfamiliar things and overestimate the risk.
3. B
4. CDFBAEG (1 mark for each of: D anywhere before F, similarly F before B, B before A, A before E, E before G.)
5. Fuel rods = high, protective clothing = low, fuel containers = intermediate.
6. (a) A nucleus splits into two parts of roughly equal size.
   (b) A million times more in a nuclear reaction than a chemical reaction.
   (c) i) T ii) T iii) F iv)T
7. Example answer: Nuclear fusion occurs when two light/small (1) nuclei get close enough and join together (1) to form a larger nucleus releasing a large amount of energy (1) in the process. The energy is given by $E = mc^2$ (where m = mass and c = speed of light), (1). The protons and neutrons in the nucleus are held together by the strong force (1) The difficulty is in getting the nuclei close enough for the strong force to take effect because the nuclei are positively charged and positive charges repel (1). The nuclei must be heated to a high temperature so they have a lot of kinetic energy (1). This is what happens in the Sun (1) where hydrogen nuclei (1) are fused to give helium nuclei (1).
   Marks will be awarded depending on the number of relevant points included in the answer and the spelling, punctuation and grammar. In this question there are about 10 relevant points so 8 or 9 points with good spelling, punctuation and grammar will gain full marks.

Chapter 8

1. D
2. ticks for: F,T,T,F,T,T
3. (a) The eye lens would get fatter.

(b) The camera lens would move further away from the point, to the left of the diagram.

4. A

5. (a) 0.5 seconds of arc

(b) A star with angle of parallax 1 second of arc is 1 parsec away. If the angle is half that the star is twice the distance away.

6. (a) 1st ray: straight line drawn with ruler from top of arrow O parallel to axis until it reaches lens, then straight line from that point through point F. 2nd ray: straight line from top of arrow O through centre of lens. Both lines extended by dotted lines back to where they cross on the left. Large upright dotted vertical arrow drawn with top at point where dotted lines cross. See 'ray diagram for a magnifying glass' on p.174 for reference.

(b) 1, 2 and 3. real, smaller than object, inverted (in any order)

7. (a) E **(b)** AB **(c)** it will be brighter.

(d) (i) C **(ii)** B

(iii) A = objective and C = eyepiece because magnification = focal length of objective divided by focal length of eyepiece so magnification = f(A) ÷ f(C) = 50 cm ÷ 2.5 cm = 20

(iv) B = objective and C = eyepiece so magnification = 25 cm ÷ 2.5 cm = 10 so this telescope has a magnification of 10 so the distant object will not be magnified as much, but because the diameter of the objective is larger (12.5 cm instead of 10 cm) the aperture is larger so the image will be brighter.

Chapter 9

1. (a) A step–up transformer has more turns in the secondary coil than the primary coil. A step-down transformer has more turns in the primary coil than the secondary coil.

(b) Mains electricity is a.c. so that the voltage can be easily changed using transformers to higher and lower voltages. Transformers will not work with d.c. The National Grid has hundreds of miles of overhead cables, and the power loss in the cables depends on the current and the resistance of the cables. By transmitting at a high voltage the same amount of power can be transmitted using a low current, because $P = IV$.

Marks will be awarded depending on the number of relevant points included in the answer and the spelling, punctuation and grammar. In this question there are about 6 relevant points so 5 or 6 points with good spelling, punctuation and grammar will gain full marks.

(c) 1.2. and 3. Increase speed of rotation, increase magnetic field strength, increase turns on coil. (any order)

2. 37°C = 310 K, absolute zero = 0 K, Freezing water = 273 K − 37°C = 236 K

3. T, F, T, F, F (not with each other, but with the walls of the container)

4. 20°C = 20 + 273 = 293K 2 × 293 = 586 K = 313 °C

5. (a) sound with frequency higher than the upper threshold of human hearing

(b) Some is reflected from each boundary between tissue layers. The reflections are used to build up an image.

(c) breaking up stones e.g. kidney stones

6. (a) They both use a gamma camera to record gamma rays/ In both cases patients are given a radioactive tracer.

(b) The PET scan looks for pairs of gamma rays that arrive at the same time in opposite directions/ The tracer is a positron emitter not a gamma emitter.

(c) By bombarding a stable isotope of an element with protons.

(d) By tagging a molecule normally used by the body with a radioactive nucleus (for example replacing an oxygen atom in glucose with a radioactive fluorine nucleus) and injecting it into the bloodstream.

(e) Looking for brain tumours/tumours/parts of the brain that are active when different activities are performed.

7. (a) X-rays are ionising radiation and could damage the cells of the fetus/ damage DNA cause cell mutation/cancer

(b) distance = ½ × 1550 m/s × 0.02 ms
= ½ × 1550 m/s × 0.00002 s
= 0.0155m (1.55 cm)

Notes

Index

Index

Periodic Table

											4 **He** helium 2

1	2											3	4	5	6	7	0
																	4 **He** helium 2
7 **Li** lithium 3	9 **Be** beryllium 4											11 **B** boron 5	12 **C** carbon 6	14 **N** nitrogen 7	16 **O** oxygen 8	19 **F** fluorine 9	20 **Ne** neon 10
23 **Na** sodium 11	24 **Mg** magnesium 12											27 **Al** aluminium 13	28 **Si** silicon 14	31 **P** phosphorus 15	32 **S** sulfur 16	35.5 **Cl** chlorine 17	40 **Ar** argon 18
39 **K** potassium 19	40 **Ca** calcium 20	45 **Sc** scandium 21	48 **Ti** titanium 22	51 **V** vanadium 23	52 **Cr** chromium 24	55 **Mn** manganese 25	56 **Fe** iron 26	59 **Co** cobalt 27	59 **Ni** nickel 28	63.5 **Cu** copper 29	65 **Zn** zinc 30	70 **Ga** gallium 31	73 **Ge** germanium 32	75 **As** arsenic 33	79 **Se** selenium 34	80 **Br** bromine 35	84 **Kr** krypton 36
85 **Rb** rubidium 37	88 **Sr** strontium 38	89 **Y** yttrium 39	91 **Zr** zirconium 40	93 **Nb** niobium 41	96 **Mo** molybdenum 42	[98] **Tc** technetium 43	101 **Ru** ruthenium 44	103 **Rh** rhodium 45	106 **Pd** palladium 46	108 **Ag** silver 47	112 **Cd** cadmium 48	115 **In** indium 49	119 **Sn** tin 50	122 **Sb** antimony 51	128 **Te** tellurium 52	127 **I** iodine 53	131 **Xe** xenon 54
133 **Cs** caesium 55	137 **Ba** barium 56	139 **La*** lanthanum 57	178 **Hf** hafnium 72	181 **Ta** tantalum 73	184 **W** tungsten 74	186 **Re** rhenium 75	190 **Os** osmium 76	192 **Ir** iridium 77	195 **Pt** platinum 78	197 **Au** gold 79	201 **Hg** mercury 80	204 **Tl** thallium 81	207 **Pb** lead 82	209 **Bi** bismuth 83	[209] **Po** polonium 84	[210] **At** astatine 85	[222] **Rn** radon 86
[223] **Fr** francium 87	[226] **Ra** radium 88	[227] **Ac*** actinium 89	[261] **Rf** rutherfordium 104	[262] **Db** dubnium 105	[266] **Sg** seaborgium 106	[264] **Bh** bohrium 107	[277] **Hs** hassium 108	[268] **Mt** meitnerium 109	[271] **Ds** darmstadtium 110	[272] **Rg** roentgenium 111							

1
H
hydrogen
1

Elements with atomic numbers 112–116 have been reported but not fully authenticated

*The lanthanoids (atomic numbers 58–71) and the actinoids (atomic numbers 90–103) have been omitted.

The relative atomic masses of copper and chlorine have not been rounded to the nearest whole number.